DISEASES

PERENNIALS

A BALL GUIDE

IDENTIFICATION AND CONTROL

A.R. Chase
Margery Daughtrey
Gary W. Simone

Ball Publishing

Batavia, Illinois, USA

Ball Publishing
335 North River Street
Batavia, Illinois 60510 USA

Library of Congress Cataloging-in-Publication Data

Chase, A. R. (Ann Renee)
    Diseases of annuals and perennials : a Ball guide : identification
and control / A.R. Chase, Margery Daughtrey, Gary W. Simone.
      p. cm.
    Includes bibliographical references (p.   ) and index.
    ISBN 1-883052-08-4
    1. Annuals — Diseases and pests — Identification. 2. Perennials —
Diseases and pests — Identification. 3. Annuals — Diseases and
pests — Control. 4. Perennials — Diseases and pests — Control.
I. Daughtrey, Margery, 1953- . II. Simone, Gary W. (Gary Wayne),
1950- . III. Title.
SB608.O7C48 1995
635.9′23 — dc20
                                           95–11629
                                              CIP

CONTENTS

ABOUT THE AUTHORS

Ann R. Chase received her doctorate in plant pathology from the University of California at Riverside in 1979. She started at the University of Florida the same year as Assistant Professor of Plant Pathology on ornamentals. Her research covered disease diagnosis, prevention, and control, with emphasis on foliage plants, cut foliage crops, and bedding plants. She has written books, chapters in books, and more than 70 refereed papers and 500 popular articles. The *Compendium of Ornamental Foliage Plant Diseases* (available from the American Phytopathological Society) and the *Ball Field Guide to Diseases of Greenhouse Ornamentals* (available from Ball Publishing), co-authored with Margery Daughtrey, have been available to ornamentals producers for many years.

When Ann retired from the Central Florida Research and Education Center in Apopka, University of Florida, in January 1994, she was named Professor Emeritus. She continues to work occasionally with the center. In February 1994, she started a writing/consulting/contract research business in California specializing in disease control on ornamental crops in the United States, Central America, and South America.

Margery Daughtrey has 17 years of experience as a plant pathologist specializing in ornamental plant diseases. She is a Senior Extension Associate with the Department of Plant Pathology of Cornell University, stationed at the Long Island Horticultural Research Laboratory in Riverhead, New York. Margery has a bachelor's degree in biology from the College of William and Mary and a master's degree in plant pathology from the University of Massachusetts. Her extension education program focuses on teaching effective disease management for greenhouse and nursery crops and landscape ornamentals.

Margery has first-hand familiarity with proper identification through the ornamental plant disease diagnostic lab she directs on Long Island. She also conducts research on disease management and has participated in many local and national programs on disease avoidance.

Gary Simone has been affiliated with the ornamental industry since he worked his way through college in retail and production nurseries in Connecticut. Gary received his Ph.D. degree in plant pathology at the University of Illinois in 1977. Currently he is Extension Plant Pathologist at the University of Florida-Gainesville concentrating on diagnosis and control of plant diseases, particularly of ornamentals and turf grass. He also coordinates the Florida Extension Plant Disease Clinic's central laboratory, which serves Florida plant producers.

Gary teaches plant disease diagnosis and researches disease diagnosis, viral diseases, and new diseases of ornamentals. He is the author of more than 450 extension and research publications and lectures at international, national, regional, and local conferences and workshops.

Introduction

Diseases of Annuals and Perennials: A Ball Guide will supply a basic understanding of the general types of diseases that can affect garden plants. Because diseases are often specific (limited to a species or a family of plants), the diversity of plants typical of a nursery or flower garden can reduce the chances of epidemic losses. Expert diagnosis and treatment of common disease problems allows best disease management to keep plants as healthy and attractive as possible. Table 1 lists scientific and common names of the annuals and perennials included in this book as well as many others.

To minimize pesticide applications, first select the most disease-resistant or tolerant cultivars available. Use good cultural practices as your second line of defense against disease (table 2). Finally, consult your local Cooperative Extension Service Office for specific recommendations of chemicals registered for control of diseases and other garden pests. Extension professionals can also recommend an extension, regulatory, or private-sector diagnostic laboratory as backup for those diseases that are particularly difficult to diagnose. For additional reading, please see the references at the end of this book.

FOLIAR DISEASES

Fungal leaf spots and blights

Leaf spots often occur on particular cultivars of annual and perennial flowers while having little or no effect on other cultivars of the same species or other plants nearby (fig. 1). This selectivity allows the horticulturist to choose the most healthy, disease-resistant variety for subsequent plantings. The term blight is used when leaf spot diseases become severe. Leaves may be completely affected and killed, with flowers and stems also invaded (fig. 2). Blight symptoms develop only when weather conditions are optimal for fungus infection and development.

Some fungal leaf spot diseases can be identified by their sporulation on infected tissues. *Botrytis* forms grayish-green or grayish-brown masses of dusty spores and is commonly called gray mold (fig. 3). Other fungi form different fruiting bodies and may look like tiny grains of pepper in the spots from where they can be scattered or form in concentric rings (fig. 4).

One of the best known fungus blight diseases on garden flowers is Botrytis blight of peony (fig. 5). New shoots may be killed back by the fungus, leaves and buds may be spotted or blighted, and large portions of open flowers may be browned. To control Botrytis blight of peony, cut stalks off at ground level in the fall and destroy them. In some cases follow this cultural control by fungicides applied to the new growth in the spring. Further, remove any blighted shoots or buds as soon as you detect them. Injury from these and other leaf spot diseases varies with annual weather conditions and by cultivar.

Leaves spotted by fungal pathogens are a source of spores that can infect new plants or new growth, and they should be destroyed. Never use them as a mulch in commercial plantings or the garden. Crowded conditions are ideal for leaf spot infections because the moisture and high humidity that result from poor air circulation between plants favor disease development. Water early in the day to allow leaves to dry thoroughly before night. Try to direct irrigation to the soil or potting medium surface rather than the leaf canopy. Drip or other ground level irrigation is best for minimizing leaf spot diseases.

In the nursery, fungal leaf diseases must be controlled during all stages of production. Examine new annuals and perennials as they come into the nursery and discard any with obvious symptoms. Leaves with spots can be

1 Coniothyrium leaf spot on *Yucca*.

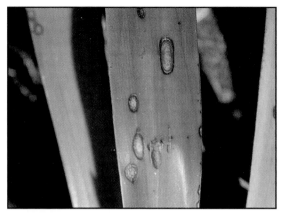

3 Sporulation of *Botrytis* on *Zinnia* exhibits powdery grayish-green spores.

2 Anthracnose disease on *Hemerocallis* (daylily) showing blight symptoms.

4 Close-up of *Septoria* fruiting bodies generating a leaf spot on *Gaillardia*.

5 Botrytis blight on *Paeonia* ◀ (peony).

removed at any time to reduce spread to adjacent healthy plants. Irrigation method, timing, and frequency are more easily controlled in commercial production systems and can be very significant in reducing severity or frequency of foliar fungal diseases.

Manipulating environmental conditions with venting and heating can give excellent control of Botrytis blight in greenhouses. At sunset, warm, moisture-laden greenhouse air promotes germination and development of *Botrytis*. To help control *Botrytis*, exchange the greenhouse air with cooler, drier outside air and then heat it. The new air will contain a minimum amount of water, and subsequent condensation onto foliage over the night is decreased.

Foliar fungicide application can supplement cultural control methods. In many cases, treatments with fungicides are best begun in the spring or on new crops as new leaves are developing. This can prevent spores that have overwintered in old leaf or stem tissue from infecting new leaves. Several fungicides are available with protectant or systemic modes of action. Protectants must be applied to the canopy before disease begins and reapplied after heavy rain or irrigation or as new growth emerges.

Systemic products are absorbed into the plant and redistributed into new growth during emergence. The newer systemic fungicides often offer an eradicant/curative feature, stopping infections within 12 to 36 hours of their initiation. Discontinue sprays in dry weather unless overhead irrigation is used frequently.

Some of the common fungal leaf diseases of annual and perennial flowers include Septoria leaf spots on *Aquilegia* (columbine), *Aster, Chrysanthemum, Coreopsis, Phlox,* and *Rudbeckia* (black-eyed Susan); Colletotrichum leaf spot on *Hosta* and *Bergenia*; Didymellina leaf spot on *Iris*; and Phyllosticta leaf spots on *Delphinium, Heuchera* (coral bells), *Iris, Liatris* (gayfeather), *Monarda* (bergamot or bee balm), and *Rudbeckia* (black-eyed Susan). Tables 3 and 4 have listings of garden flowers affected by *Corynespora cassiicola* and *Myrothecium roridum*, respectively. Both of these diseases are most often seen in the hot, humid areas of the Southeast.

Bacterial leaf spots and blights

Leaf spots caused by bacteria cannot always be easily distinguished from fungal leaf spots by visual examination. In most cases, bacterial diseases are aggressive in hot weather because high temperatures favor growth of

6 Mixed infection of *Alternaria* (fungus) and *Xanthomonas* (bacterium) on *Zinnia.*

7 Bacterial leaf spot on *Clivia* displays water-soaking.

the bacteria. Generally, fungi are most active under somewhat cooler conditions. However, some plants, such as *Zinnia,* are commonly infected by both fungal and bacterial pathogens (fig. 6). This makes diagnosis and treatment more complicated.

Frequently bacterial leaf spots are characterized by water-soaking around dead tissue in the leaf (fig. 7). This symptom is most easily seen in the morning while dew remains on the plant. At other times a yellow halo forms around the spot (fig. 8). Bacteria do not produce spores like most fungi, instead bacterial cells are spread by splashing water off the infected plant surface. Overhead or sprinkler irrigation aggravates bacterial and fungal leaf spots and blights. To control this disease class, minimize overhead irrigation.

8 Sometimes a yellow halo forms around bacterial spots, such as these on *Delphinium*.

9 Yellow spots appear on upper leaf surfaces of *Antirrhinum* (snapdragon) infected with rust.

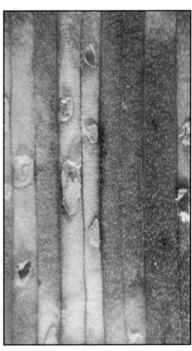

10 Close-up of rust pustules on the underside of *Iris* leaf.

Annuals and perennials do not ordinarily develop severe bacterial disease problems. When such a condition does occur, sanitation is the best action. Removing infected leaves as soon as they are noticed is the most direct control method and generally works in a garden. In commercial production, the best bacterial disease control method is to use pathogen-free plants and rigorous scouting to identify and remove infected plants in the early stages. Always examine new plugs or cuttings for symptoms, and discard them before the disease can spread throughout the nursery. In general, chemical control of bacterial diseases is rarely successful because few bactericides are available and those that are do not work well.

Some of the bacterial disease organisms most commonly encountered are *Pseudomonas delphinii* (on *Delphinium* and *Aconitum*), *Xanthomonas campestris* pv. *tardicrescens* (on *Iris*), *Xanthomonas campestris* pv. *papavericola* [on *Papaver* (poppy)], *Xanthomonas campestris* pv. *pelargonii* [on *Pelargonium* (geranium)], *Xanthomonas campestris* pv. *zinniae* (on *Zinnia*), and *Xanthomonas campestris* pv. *begoniae* (on *Begonia*). Tables 5, 6, and 7 list garden flower hosts of *Erwinia* spp., *Pseudomonas cichorii,* and *Pseudomonas syringae*, respectively.

Rust diseases

Rust fungi usually appear first as pale spots on the upper leaf surface with pustules containing the powdery spores on both leaf surfaces (fig. 9). These spores are often a rusty reddish brown, thus, the common name "rust" for this group of diseases. The signs of rust infection can be slightly different on each plant, but, in general, they are easy to identify because the spores (white, yellow, orange, brown, red, or black) are produced in discrete masses visible on both leaf surfaces (fig. 10). Most rust fungi produce highly colored spore masses in pustules on leaves. To diagnose rust, rub infected leaves on a sheet of white paper. The resulting colored spore streaks are diagnostic for rust.

Rust fungi are highly specialized and affect only closely related plants. Some rust fungi, however, spend a portion of their life cycle on a taxonomically unrelated plant that grows in the same ecosystem. For example, one rust fungus infects only goldenrod and pine species, and the spores produced on either plant are capable of infecting the alternate host. Control this type of rust by eliminating the alternate host. However, generally such action is not practical.

In other cases, a related plant may contribute to disease development on a desirable plant. The common weed mallow (*Malva neglecta* [cheeses]), for example, may be a source of rust spores for hollyhocks or mallows (fig. 11). Eliminating the weed host in this case helps reduce rust infections. To eliminate overwintering spores, remove infected leaves so that new growth or new plants will not have a nearby source of infection. If irrigating overhead, water plants early in the day to allow leaves to dry quickly and reduce severity of most leaf diseases. Remove severely infected plants to reduce spread to adjacent healthy plants. Since rust spores are dry and easily dislodged, take care when removing rust-infected plants or plant parts. It's a good idea to carry a zip-lock bag for any infected leaves or plants while inspecting your crop.

Chemical control of rust diseases is possible for both commercial producers and the gardener, although many more chemicals are available to producers. Read labels carefully to determine which product is best for each situation, and be sure to follow directions when using a pesticide.

Many rust fungi cause little damage to the plants they infect; rust caused by *Puccinia iridis*, however, may be very damaging to certain irises. *Campanula* (bluebells) also may be quite sensitive to rust infection; leaves may be scorched and plants stunted. Rust diseases on *Malva* (hollyhock), *Antirrhinum* (snapdragon), and *Dianthus* are quite common.

Powdery mildew diseases

Powdery mildew is probably one of the most troublesome diseases of annuals and perennials because the symptoms are obvious and unsightly. Plants are rarely killed by a powdery mildew infection, although lower leaves may drop under high disease pressure (fig. 12). Powdery mildew diseases look very much the same regardless of the plant affected. Leaves develop patches of frosty white fungus growth, primarily on the upper leaf surface (fig. 13). In some cases, the leaf tissue beneath the spot becomes chlorotic (yellow) or even necrotic (brown and dead). Lower leaf surfaces, stems, pedicels, and flowers are attacked by the fungus in severe infections. Occasionally, as a response to mildew infection (fig. 14), plants develop purplish discolorations instead of the typical white spots.

Although the disease appears identical on each plant host, different species of powdery mildew fungi are responsible for infections on different plants. Mildew on roses, for example, is not a source of infection for mildew

11 Hollyhock rust on alternate host, common mallow (cheeses).

13 Close-up of powdery mildew on poinsettia showing separate spots.

12 Severe powdery mildew infections cause leaf drop such as this seen on *Monarda*.

14 Purple discoloration of *Phlox* with powdery mildew infection.

on any herbaceous perennials. The host specificity of powdery mildew provides assurance that the disease will not be epidemic unless only one plant type is being grown. All powdery mildew diseases respond similarly to environmental factors and fungicidal controls.

Powdery mildew is somewhat unusual as a foliar disease since symptoms do not begin to appear during the rainy days of spring that are so encouraging to most fungus infections. Instead, powdery mildew symptoms occur later in the year whenever drier conditions prevail. The time of occurrence is related in part to the moisture requirements of the fungi. Powdery mildew fungi sporulate when the humidity is high rather than when leaves are wet.

Conditions that favor powdery mildew are similar, and we can generalize about their control. Environmental monitoring, humidity control, and scouting are important parts of an integrated approach to powdery mildew control. Fungicides can also be used in an integrated control program.

Relative humidity can greatly influence the powdery mildew development. In most cases, free water on the leaf surface will reduce germination of powdery mildew conidia. Some vegetable greenhouses syringe or spray the crops with water to discourage conidial germination. This method of powdery mildew control also washes conidia off of the leaf surface and into the soil where they cannot infect the crop. Unfortunately, using this control on most floral crops is not advised since other diseases may be pre-

15 Cultivars exhibit dramatic differences in their response to powdery mildew infection.

sent. Diseases such as *Botrytis cinerea* or bacterial diseases as *Xanthomonas campestris* are worsened when water is applied to leaves.

To control mildew infections to a limited degree, maintain adequate aeration within a planting, and thin stands if they become too dense. When possible, choose sunny rather than shady sites and use resistant cultivars of mildew-prone annuals and perennials (fig. 15). Frequent overhead irrigation, especially late in the day or at night, keeps the humidity around plants high and promotes powdery mildew and other foliar diseases. Trellis *Lathyrus* (sweet peas) and certain other vining plants that are susceptible to powdery mildew for better aeration and higher fungicide efficacy.

One of the best ways to minimize losses from any disease is to use resistant cultivars. This allows the crop to be produced in the presence of the pathogen with minimal loss and minimal chemical application. Many growers know which cultivars of their crops are more resistant to disease, but widespread dissemination of this type of information is not common.

To control powdery mildew, follow the same sanitation steps as discussed for rust diseases: Remove infected plant material and treat with a fungicide if needed. As an additional control, reduce greenhouse humidity as described for foliar fungal disease. Several fungicides are also available to retard powdery mildew development. Protectant-type sprays require

16 Lower leaf surface of *Pelargonium* (geranium) infected with downy mildew fungus.

17 Upper leaf surface of *Pelargonium* (geranium) infected with downy mildew fungus.

frequent reapplication, since each new leaf must be treated as soon as possible for best protection. In contrast, the newer systemic fungicides allow an extended interval between applications, since their active ingredient can move into new leaves. Many of the most popular annuals and perennials are highly prone to mildew infections: *Aster, Coreopsis, Lathyrus* (sweet pea), *Monarda* (bergamot or bee balm), *Phlox, Rosa* (rose), and *Rudbeckia* (black-eyed Susan), to name a few.

Downy mildew diseases

Downy mildew diseases produce their characteristic symptoms on the undersides of leaves (fig. 16). Pale green or yellow patches appear on the upper surface of affected leaves (fig. 17), while sporulation of the fungus is seen as fuzzy (downy) white, tan, or gray areas on the leaf underside, opposite the pale spots. The older leaves are affected first and often turn completely brown and dry.

Downy mildew may be carried through the seed. Closely related weeds may also be a source of spores for an infection. *Geum* and *Potentilla* are affected by *Peronospora potentillae*, a downy mildew that is also a disease of the Indian strawberry weed (*Duchesnea indica*). *Viola* (pansy) may be affected by three different types of downy mildew. The most common downy mildew disease is found on roses in all types of commercial production and gardens all over the world.

The main control of downy mildew in gardens is to remove infected plants. Spores in the soil may allow the disease to persist for several years, so replacing plants with a different species is advisable when downy mildew has been a serious problem in previous seasons. Weed control, spacing, and watering practices that minimize the time that the foliage is wet are helpful.

In the nursery, downy mildew is often controlled with fungicides. Since fungicide resistance is known to occur with the downy mildew fungi, combination products are recommended. Some garden flowers subject to a downy mildew disease include *Artemisia, Aster, Centaurea* (bachelor's button or cornflower), *Pelargonium* (geranium), *Geum, Lupinus* (lupine), *Potentilla, Rosa* (rose), *Rudbeckia* (black-eyed Susan), *Veronica* and *Viola* (pansy).

ROOT AND STEM DISEASES

Root and stem rot

Annuals and perennials are subject to various root and stem diseases (fig. 18). An unhealthy root system usually produces symptoms of wilting, stunting, or nutrient deficiency in the leaves that may easily distract the observer from the actual source of the problem. Many of the same fungi that commonly cause damping-off on seeds and seedlings are capable of causing infections of the roots and lower stems of mature plants. These fungi are generally soilborne organisms. *Rhizoctonia, Sclerotinia, Sclerotium, Phytophthora,* and *Pythium* are five common soilborne fungi that may be present in field or garden soil or may be contaminants of a container mix. These fungi have broad host ranges, meaning that they are not associated with specific plants but affect a wide range of different species.

18 Severe symptoms of root and stem disease on *Campanula*.

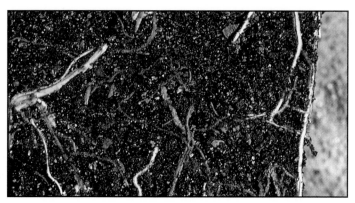

19 Root system with brown (dead) and white (healthy) roots.

Pythium and *Phytophthora* diseases are particularly favored by cool, wet soils with poor drainage. Quite often, the first observed symptom of *Pythium* infection is yellowing of the lower leaves of the plant, which is a typical indication of poor root health. Root infection takes place at the root tip but rapidly spreads throughout the root system, turning the roots gray, brown, or black (fig. 19).

Plants whose roots have been injured by excess soluble salts from overfertilization or salt water intrusion are especially susceptible to *Pythium* infection. When roots are attacked by *Pythium*, the fungus feeds on the softer outer root cortex, leaving behind only the tough vascular tissue (stele) at the root core. The root's outer portion easily sloughs off when an infected plant is pulled from the soil. Thin, stringy root cores may be all that remains of the root system (fig. 20).

20 Sloughing of outer root tissue leaving the core.

21 Pythium root rot on Gerbera daisy.

Roots infected by *Pythium* (as well as many other root pathogens) are gray or brown, in contrast to the healthy white color of actively growing roots on most plants. This discoloration is particularly easy to see on plants grown in soilless media: Simply knock the plant out of the pot and examine the roots (fig. 21). Often *Phytophthora* moves up into the stem to cause a canker at its base (fig. 22). Sometimes *Pythium* also causes a base canker on plants such as geranium.

Rhizoctonia may cause symptoms similar to *Pythium* above ground, because plants wilt and can show nutrient deficiency symptoms when they are girdled due to cankers at the soil line (fig. 23). Ordinarily roots are not affected much by *Rhizoctonia*; the fungus primarily grows across

22 Basal stem rot produced by *Phytophthora* on *Dianthus.*

23 Nutritional deficiency symptoms on
Petunia caused by Rhizoctonia stem rot.

the soil surface and attacks plant stems (fig. 24). *Rhizoctonia* often causes a symptom that is called "wire stem" by vegetable growers: The invaded portion of the stem, just at the soil surface, turns brown and shrivels, giving it a tough, wiry appearance.

This fungus grows most rapidly under relatively high temperature conditions, so losses to *Rhizoctonia* commonly occur in the hot summer months. Improving drainage will not alleviate the problem, for *Rhizoctonia* does not require free water for its growth or development.

25 Close-up of a sclerotium on vinca infected with *Sclerotinia sclerotiorum*.

24 Basal canker on poinsettia from *Rhizoctonia*.

The fungus *Sclerotinia* develops distinct black lumps of fungus tissue called sclerotia (singular, sclerotium) in the pith (inside of stem) or on the stem surface (fig. 25). These are large enough to be visible to the naked eye and are roughly round or elongated. The cottony white fungus itself may be evident at the plant base.

The soil pathogen, *Sclerotium rolfsii*, causes a widespread disease known as southern blight across a broad range of agronomic, fruit, ornamental, and vegetable crops. This fungus prefers high temperatures (84 to 90F/29 to 32C) and high soil moisture. Diseases produced include damping-off, root and crown rot, and fruit decay. Affected plants exhibit a wilt or decline that may be quite rapid (a few days), followed by plant death. During moist weather, the fungus may be observed as coarse, white mycelial mats that emerge at the soil line and often grow up the plant stem or over adjacent soil (fig. 26).

The overseasoning, vegetative, seed-like propagule called a sclerotium is round, white, and the size of a mustard seed. It can mature in days to a tan

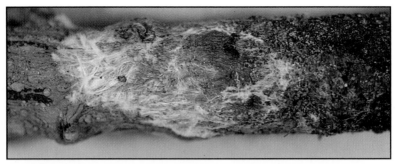

26 White fan-like growth of *Sclerotium rolfsii* on a *Capsicum* (pepper).

27 Sclerotia of *Sclerotium rolfsii* on a *Helianthus* (sunflower) stem base.

color and fall to the soil surface (fig. 27). This fungus spreads by movement of infested soil during hand or mechanical cultivation operations. Dig infected plants carefully, removing them with adjacent soil and destroying them. Fungicide applications are not particularly useful against this pathogen in the landscape. See table 8 for a list of annuals and perennials reported as hosts of *Sclerotium rolfsii*.

Root- and stem-rotting fungi are capable of infecting many plants, so removing infected plants after a disease outbreak is very important. Remove wilted plants (and some of the adjacent soil) immediately when plants are found to have a rotted root system or basal cankers. Plants infected by soil fungi that produce sclerotia (*Rhizoctonia, Sclerotinia,* and *Sclerotium*) must be dug carefully to avoid scattering the highly persistent, seedlike structures and minimize the chance that the fungus will spread to nearby plants. An appropriate fungicide can be drenched at the site where a plant has died. In the nursery, discard infected plants and use a fungicide drench on nearby plants of the same species. Be sure to check with local state officials for lists of fungicides that are legal and safe for use.

In commercial production, the first step to disease control is using healthy plugs or cuttings for propagation. A well-drained potting medium with balanced nutrition will minimize plant stress. Overwatering plants creates ideal conditions for disease development and spread.

It is a good idea to grow plants on raised benches when possible to reduce infection from the native soil. Wash and sterilize pots, trays, and benches between crops. Discard pots and trays from infected crops. Preventative treatments with soil drench fungicides are often used, but not generally recommended. Most of the fungicides can cause slower growth on small seedlings and should not be used routinely.

Seedling diseases (damping-off)

Damping-off refers to the collapse of young seedlings caused by infection with a plant pathogen (usually a fungus). Infection may occur either before or after seedlings emerge. If it occurs before emergence, damping-off is easily confused with poor seed germination. Seedlings may also be killed after emergence when either the roots or the stems are attacked (fig. 28). The general pattern of damping-off in a broadcast-seeded tray is a circular patch of toppled seedlings. Plug seedling trays generally restrict disease spread within the tray, since the walls between the cells slow fungus movement into adjacent cells (fig. 29).

Although many different species of fungi may cause damping-off, *Rhizoctonia* is the one most often responsible for this disease. *Rhizoctonia* is common in natural soil, but steam-pasteurization or chemical fumigation destroys *Rhizoctonia* and other harmful fungi and renders soil safe for use in a germination mix. Soilless mixes containing peat, perlite, and vermiculite are popular with commercial producers and are largely free from pathogens that could cause damping-off diseases as well as other pests, such as weed seeds and insects. In addition, many soilless mixes promote excellent seed germination and growth by providing optimal moisture as well as good drainage.

Soil-based or soilless media can be sterilized with either dry heat in a traditional oven or by radiation in a microwave oven. Small volumes of soil can be effectively pasteurized when heated to 160 to 180F (71 to 82C) for at least 30 minutes. Soil should be moist and media placed in a glass baking

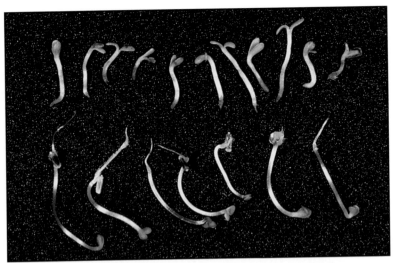

28 Root and basal stem infection on *Chrysanthemum* (Shasta daisy) seedlings (top row).

29 Pythium damping-off of *Browallia* seedlings in a plug tray.

dish for heating. Similarly, the microwave oven can be used for pasteurization of soil in plastic freezer bags. This technique's effectiveness increases with duration of exposure in the microwave, but decreases when too much soil is treated at once. High soil moisture also decreases effectiveness. The optimum procedure is to heat about 2 to 2.5 pounds (1 kg) of slightly moist soil for at least three minutes at full power. The treated soil can be stored in the same container.

Additionally, maintain clean working conditions to minimize *Rhizoctonia*. Keep potting mix away from direct contact with the ground to avoid contamination. Clean and disinfect mix storage, potting, and bench surfaces. Regular household chlorine bleach (5.25% sodium hypochlorite) can be used at a 1:9 dilution (1 part bleach to 9 parts water). Wooden benches can be treated periodically with copper naphthenate. Remember to aerate copper napthenate-treated flats for at least a week before use. Skipping any sanitation step may result in costly losses.

Grow seedlings on raised benches if possible or on clean black plastic sheets or plastic weed matting if plants must be placed on the ground. Since some states restrict use of copper naphthenate and sodium hypochlorite, check with local officials to verify conditions under which you can use these chemical compounds.

Pythium, Botrytis, and *Alternaria* can also cause damping-off diseases. *Pythium* is a water-mold fungus that prospers best in a perpetually soggy, poorly drained growing mix. *Pythium* is one of the reasons why it is important to have adequate air pore space in a germination mix. Wet soils favor *Pythium* attack, and seeds may be rotted as they start to germinate or wilt and die when their roots are infected.

Botrytis is best known for attacks on flowers or on foliage of nursery stock during overwintering, but it can also cause damping-off. *Botrytis* infection can occur when greenhouse humidity is too high and condensation forms on seedlings. Free moisture on plant surfaces during the night is also favorable for *Botrytis* infection.

Control air circulation to reduce humidity, and keep the greenhouse free from plant debris to reduce inoculum for *Botrytis* infections. During winter vent greenhouses late in the afternoon to reduce relative humidity below 85% and lower the likelihood that water will condense on plants overnight.

Alternaria can also cause damping-off (fig. 30). This fungus may sometimes be carried on the seed itself and because of that, some commercial seed companies may treat certain seed lots prior to packaging and sale. Seed treatment with a fungicide can be noted from the yellow, red, blue, or green stain on the seed. This is an adequate preventative control measure for damping-off when it is available.

If treated seed is not available, you can treat seed at home. Seed treatment with a solution of bleach (1 part bleach to 5 parts water) for 15 to 30 minutes can effectively kill most fungal spores on the outside of seed. Be sure to test a small batch of seed for sensitivity to bleach. Carefully count

30 Alternaria damping-off on *Eustoma* (lisianthus).

two small batches of seed; 25 to 100 are sufficient depending upon seed cost and availability.

Soak one batch in the test solution and the other batch in plain water. Plant both batches, being sure to label them carefully for later examination. As seedlings emerge you can count them. It is sometimes helpful to keep a record of the dates they emerge to note any change in the number of days required for emergence.

Sometimes seed treatments improve germination, so don't be surprised if more of the treated seed grow than the water-soaked seed. If seed germination is reduced by the bleach treatment, you can lower the time of exposure or the concentration of the bleach solution and perform the test again. Plant treated seed immediately, or dry them under low heat (a 250F [120C] oven should be safe for most seed) if you must store them for short periods of time. Do not use a microwave since this will kill the seed. Some seed are sensitive to cold, so they should be stored at room temperature under dry conditions. This is especially true of tropical and subtropical plants.

Other cultural controls can minimize damping-off diseases. When watering seed flats by hand, avoid splashing and use only a clean hose nozzle that is not allowed to drop on the ground. Properly regulating moisture, light, temperature, and nutrients, according to each plant's requirements, produces vigorous seedlings that emerge quickly and are more resistant to the fungi that cause damping-off diseases. When direct-seeding into the ground, sow seed only when the temperature and soil moisture conditions

31 Close-up of a crown gall on chrysanthemum cuttings.

are optimal for rapid germination. Sowing early into cold soil can slow germination and increase seed vulnerability to damping-off diseases.

Crown gall disease

Crown gall disease is caused by a soil bacterium (*Agrobacterium tumefaciens*) and is known to occur on plants in at least 93 families. Diseased plants are identified by the presence of galls, usually at the stem base (fig. 31). Sometimes galls form well above the base of the plant, but ordinarily the only aboveground symptom is stunting. Galls on roots or stems disrupt the normal flow of water and nutrients and lead to reduced growth. Crown gall can also increase sensitivity to winter injury on some plants.

Roses are some of the plants most notoriously susceptible to crown gall injury (fig. 32). Damage on most hosts is infrequent and minimal. Accurate laboratory diagnosis is recommended, since other bacteria, fungi, nematodes, insects, and mites can also cause galls.

Crown gall is caused when the bacterium attaches to a wounded area of the plant, to the crown, or anywhere along the stem where roots emerge. The wound allows transmission of plasmids (mobile genetic elements composed of tiny circles of deoxyribonucleic acid [DNA]) from the bacterium into the plant cells. Once inside, the plasmid causes the plant cells to multiply rapidly and form a characteristic gall.

32 Crown gall at graft union on a potted rose.

Crown gall bacteria are spread through movement of infested soil and plants. Carefully remove infected plants as soon as they are detected, along with the soil in close contact with their roots. Replant the area with a non-host or a less susceptible species if one is known. Crown gall is only rarely encountered on annuals and perennials. Among the reported hosts for the disease are *Achillea* (yarrow), *Aster, Chrysanthemum, Dahlia, Delphinium, Phlox,* and *Rosa* (rose) (table 9).

Chemical controls for this disease have not been successful, although biological controls have had good success in reducing crown gall infections. A beneficial bacterium, *Agrobacterium radiobacter* K-84, is used commercially on roses as a preventive treatment for bare root nursery stock before planting. The K-84 bacteria produce an antibiotic that is toxic to some strains of the crown gall bacterium.

Canker and dieback diseases

Cankers are simply dead portions of plant stem tissue, just as leaf spots are dead tissue on leaves. When this dead portion of the stem is at the end of the branch or shoot, it is often referred to as dieback. Cankers are brown or black areas that may become shrunken with time as the healthy adjacent tissue continues to grow. The extent of a canker's growth is often a matter of the plant's condition. For example, a plant under drought stress will often develop more extensive cankers than a nonstressed plant.

33 Dieback on rose caused by
Coniothyrium and other fungi.

Bacteria or fungi that cause cankers often enter the stem through wounds made by hail, insects, or pruning wounds. Another entry route is through natural openings, such as lenticels in stems. Sometimes a fungus or bacterium enters through stomates in leaves or nectarthodes (nectar glands) in flowers and grows into adjacent stem tissue. When conditions are favorable, the pathogen grows extensively and can girdle branches, shoots, or main stems causing wilting and even death.

The most common bacterial canker is fire blight caused by *Erwinia amylovora*. Fire blight is common on pear, apple, hawthorne, pyracantha, raspberry, and cotoneaster. Most cankers seen in the garden, however, are caused by fungi. Roses may be plagued by fungal cankers caused by *Botrytis, Coniothyrium,* or *Phomopsis* (fig. 33). *Rhododendron* may wilt and die back when cankers caused by *Botryosphaeria* attack plants stressed by drought or those planted too late in the fall.

Carefully prune out cankers found on branches of annuals or perennials to reduce further spread of the disease in those plants as well as to remove a source of disease for other plants. Always prune when the plant surfaces are dry to minimize the chance of spreading the disease, since most of these pathogens require free water to infect even wounded surfaces. Make each cut well below discolored areas, cutting just above a node at a slight slant in order not to leave a large stub. These stubs usually die back and are, unfortunately, perfect sites for new cankers or diebacks to start.

When pruning to control a bacterial disease such as fire blight, take extra precautions. Prune plants when the disease is not active, usually the win-

34 Fusarium wilt of *Dianthus*.

ter months, and use 70% alcohol to clean the pruners between each cut. Make cuts 12 inches (30 cm) below the cankered area, and remove and burn the pruned stems.

To minimize most canker and dieback diseases, employ the following preventive maintenance steps. Keep plants well watered during periods of drought, but avoid wetting the foliage for extended periods of time. Drip watering systems will accomplish both with the greatest ease. Select disease-resistant cultivars whenever possible.

Vascular wilt diseases

Vascular wilt, caused by fungi or more rarely bacteria, systemically invade a plant's vascular tissue (xylem or phloem). The most obvious symptom of vascular wilt is water stress or wilt that can readily kill sections of the plant (fig. 34). Often the aboveground symptoms of vascular wilt are similar to those caused by root rots. Both types of disease classes cause water stress, even though the roots themselves may look healthy. Plants may be stunted, have yellow or scorched leaves, and ultimately wilt and die. Because some vascular wilt pathogens move upward from the roots to the leaves within the xylem, symptoms sometimes appear on only one side of the plant or even on only half of a leaf (fig. 35). A cross section of a stem with infected xylem tissue reveals dark dots of discoloration that are seen as dark streaks when the stem is cut lengthwise (fig. 36).

The fungi most often responsible for vascular wilts of annuals and perennials are *Fusarium oxysporum*, *Verticillium albo-atrum*, and *Verticillium*

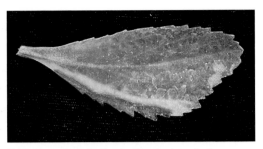

35 One-sided death of a Shasta daisy leaf infected with Acremonium wilt.

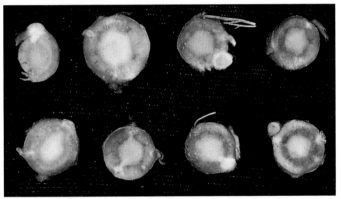

36 Shasta daisy stem infected with Acremonium wilt reveals vascular discoloration.

dahliae. Fusarium fungi tend to be very specific, attacking only plants that have a close taxonomic relationship. In contrast, *Verticillium* fungi attack many different plants, although plant species have varying susceptibility to the pathogen. Verticillium wilt occurs occasionally on plants such as *Aconitum, Aster, Chrysanthemum, Coreopsis, Dahlia, Delphinium, Dicentra* (bleeding-heart), *Impatiens, Paeonia* (peony), *Papaver* (poppy), and *Phlox.* Fusarium wilt is most often a problem on *Callistephus* (China aster), *Chrysanthemum, Dahlia* and *Dianthus barbatus* (sweet William).

A vascular wilt disease is often unknowingly introduced to a garden or a nursery because plants can be systemically infected and yet not show symptoms until they are subjected to high temperature, drought, or some other stress. These diseases must be controlled by removing affected plants. It is critical for plant propagators to avoid using any portion of an infected plant, since all portions may be infected whether they show symptoms or not. Always use pathogen-free indexed stock to ensure a good crop. Fumigate ground beds to reduce contamination in production areas.

One bacterial pathogen, *Pseudomonas solanacearum*, causes a widespread vascular wilt disease in the warmer climate zones of the United States. This bacterium invades plant roots through natural wounds produced during root emergence or through root injuries caused by close cultivation. Many plants in the Asteraceae and Solanaceae families are attacked; the disease is particularly severe in ornamental *Capsicum* spp. (peppers), *Dahlia*, *Helianthus*, *Tagetes* (marigold), and *Zinnia* spp.

Affected plants develop a progressive wilt that begins at midday, lasting for longer periods each day until the plant dies. Roots are often decayed, with a dark brown discoloration visible in the pith (inner tissue) of the lower stem.

One diagnostic test for bacterial wilt disease is the streaming test. Cut a 2- to 3-inch (5 to 7 cm) section from the lower stem (starting at the soil line). Hold this section in a clear glass of water and look through the glass at a light source. If bacterial wilt is affecting that plant, thin white streams of bacteria (like smoke tendrils) will emerge from the cut end of the stem (fig. 37).

Pseudomonas passes through ground beds with the movement of infested soil during cultivation or through surface water movement. Rotating to a nonsusceptible plant species is an excellent control for an infested area. Soil solarization of infested areas can aid in managing bacterial wilt diseases in certain parts of the United States. Soil fumigants can be useful in the short term, but these products are highly toxic and not legally available to most gardeners. Commercial producers can more easily employ fumigants for treatment of potting medium and rooting beds.

VIRUS AND MYCOPLASMA DISEASES

Virus diseases

Viruses are particles of protein and genetic material (RNA or DNA) so small that they can only be seen with the aid of an electron microscope. One of the most common ways for a virus disease to spread is by using infected stock plants for propagation. Viruses are also frequently carried or vectored from plant to plant by aphids or other piercing, sucking insects, such as thrips, planthoppers, and leafhoppers.

Some viruses have a large host range, whereas others are currently

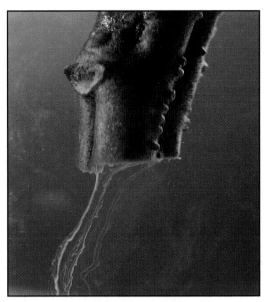

37 Bacterial streaming from cut ends of stems is diagnostic for bacterial wilt disease.

thought to have a single plant host. Although a virus may be named after one of its plant hosts, many other plants may also be susceptible. Cucumber mosaic virus, for example, infects many other species besides cucumber (table 10).

Symptoms of virus infection are often easy to distinguish from symptoms of other contagious diseases, such as those caused by bacteria or fungi. However, a virus infection can be difficult to distinguish from other problems, such as herbicide injury or nutrient imbalance. The most common symptom of virus disease is stunting. Stunting is not always recognized since often all of the plants are infected and therefore grow similarly. Without a healthy, uninfected plant for comparison, growth reduction is hard to see.

Other symptoms that are more commonly recognized as indications of virus infection are mosaics (fig. 38), mottles or ringspots (fig. 39), and color breaks (fig. 38). These patterns stand out on the leaf since affected areas are either lighter or darker than the normal leaf color. Distortions are also common symptoms of virus infection that can be confused with 2,4-D damage, injury from another growth-regulating herbicide, or from insect/mite feeding. Distortion is characterized by cupped, strapped, or unusually small leaves (fig. 40).

38 Mosaic caused by a virus on *Zinnia*.

39 Jagged patterns on *Dicentra* are induced by a virus.

40 A virus on *Vinca* causes leaf distortion and mosaic.

Once a plant is infected with a virus, any new plants vegetatively propagated from it will probably carry the same virus. Virus diseases therefore must be controlled by using only healthy plants for propagation. Virus-indexing is commercially used to ensure pathogen-free plants. Be sure to purchase plants from only reputable sources. Maximize the value of using indexed plants by using new or sterilized potting medium. For future propagation of virus-free cuttings, a mother block can be established from indexed stock.

In addition, insects must be controlled since some of the most common viruses are spread by insects. Using labeled insecticides to control insect vector populations may also reduce incidence of virus disease. Cucumber mosaic virus is vectored by certain aphids, while tomato spotted wilt virus is vectored by certain thrips. Since some viruses are common on weeds, the weeds must be controlled as well in order to eliminate them as a potential source of either the virus, the insect vector, or both.

Some of the common viruses on annuals and perennials are the peony ringspot virus on *Paeonia* (peony); cucumber mosaic virus on *Aquilegia* (columbine), *Dahlia*, and *Delphinium*; iris mosaic on *Iris*; and impatiens necrotic spot or tomato spotted wilt on *Aster, Catharanthus, Dahlia, Delphinium, Gaillardia* (blanket flower), *Impatiens*, and others.

Aster yellows disease

Aster yellows disease is caused by a mycoplasma-like organism (MLO). MLOs are similar to bacteria in their shape and structure, but they create virus-like symptoms in the plants they infect. The aster yellows MLO is vectored by *Macrosteles fascifrons*, the aster leafhopper. Symptoms of aster yellows are often dramatic yellowing and proliferation of adventitious buds into bushy witches'-broom (fig. 41). Flowers show the most obvious symptoms, including reduced size and petals that are partially or wholly green (fig. 42).

To control aster yellows in nurseries, exclude the leafhopper vector from the growing area by using tightly woven netting, since insecticides do not provide sufficient control. The aster yellows MLO is not seedborne, but it has many weed hosts; leafhoppers that have previously fed on weeds can easily introduce aster yellows disease to annuals and perennials.

Even though aster yellows disease is rare, you should remove infected plants immediately, so that leafhoppers feeding on them do not spread the infection further. Annuals that may be infected include *Callistephus, Catharanthus, Petunia, Tagetes* (marigold), and *Viola* (pansy). Perennials such as *Bellis, Campanula, Chrysanthemum, Coreopsis, Delphinium, Gaillardia, Rudbeckia,* and *Salvia* are also susceptible to aster yellows.

Nematode diseases

Nematodes are microscopic, unsegmented worms. Most nematodes in the soil feed on bacteria, fungi, and other soil microorganisms, while a few specialized species feed on plants. Most plant parasitic nematodes feed on

41 Witches'-broom symptoms are created by the aster yellows mycoplasma on vinca.

42 Aster yellows infection on *Coreopsis* shows distorted flowers.

plant roots. Symptoms of nematode infestation, usually stunting or wilting, are common signs that the root system is compromised (fig. 43). The most easily diagnosed symptom of nematode infestation is galls on the roots caused by the root knot nematode, *Meloidogyne* (fig. 44). Other nematodes that feed on roots include *Pratylenchus* (lesion nematode), *Ditylenchus* (stem and bulb nematode), *Radopholus* (burrowing nematode), and *Criconema* (ring nematode).

43 Stunting and wilting on *Coleus* caused by root knot nematodes.

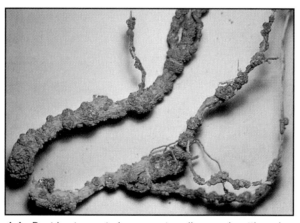

44 Root knot nematodes generate galls on poinsettia and many other plants.

Nematode control can be difficult since most infestations go unnoticed until symptoms are severe and cultural controls ineffective. Chemicals for controlling nematodes are highly toxic to pets, people, and even the plants they are designed to treat. The best control is to avoid the problem. Examine plants carefully when they are purchased to verify that they are free of obvious symptoms of nematode infestation. Check the root systems to make sure they are free of galls and appear vigorous and healthy. Examine dying plants for symptoms of root damage and, if nematodes are suspected, remove and discard all roots and their adjacent soil.

In commercial production, treat soil with a nematicide or broad-spectrum soil fumigant before replanting since many of the plant parasitic nematodes have a wide host range and must be eradicated before a new crop is planted. Rotating planting sites for susceptible plants (when possible) may help avert or delay the onset of nematode injury. Additionally, adding organic matter to sandy soils can lessen injury from root knot nematodes (*Meloidogyne*).

Foliar nematodes (*Aphelenchoides*) feed within plant leaves and stems. Leaf symptoms include wedge-shaped dead areas bordered by the leaf veins (fig. 45). Initially these areas appear reddish or yellowish, but they turn brown as they die. Some of the most commonly affected garden flowers are *Aquilegia* (columbine), *Bergenia, Chrysanthemum,* and *Heuchera* (coral bells). To control foliar nematodes, remove symptomatic leaves as soon as they are noticed to reduce spread within the plant or to adjacent plants. Sometimes it is advisable to remove the entire plant.

MISCELLANEOUS PROBLEMS THAT MIMIC DISEASES

Sometimes you will discover unusual symptoms and wonder if they are cause for concern because it seems the more diseases you know about, the more you find. The following section identifies some of the problems you may discover that are not caused by a plant pathogen.

Plants develop dramatic wilting in response to drought, and this can be confused with wilting caused by diseases. The wilt disappears when adequate water is supplied, which will not happen if the wilt is due to a disease. Marginal burning can also be a result of temporary water stress (fig. 46).

Lack of water for too long a period can cause permanent wilt, especially for pot plants; plants eventually die regardless of how much water is applied after such a severe wilt. Root rot fungi and nematodes can cause similar symptoms, since wilting due to lack of water is the primary symptom of an inadequate root system.

Excessive soil moisture can also be damaging in planting beds. Prolonged periods of saturated soil moisture caused by a stalled weather front or a poorly drained site can suffocate roots. Too much soil moisture cuts off oxygen to the roots. If the situation persists, roots will die and

45 Foliar nematode infestation on *Aquilegia* (columbine) shows wedge-shaped spots.

46 Lack of water caused marginal burning on *Tricyrtis* (toad lily).

47 Oedema on *Pelargonium* (geranium) comes from excess water ◀ in leaves.

decay—causing plant wilt and decline and predisposing the plant to secondary invasion by soil fungi.

A second adverse impact of excessive water is a physiological injury known as oedema (edema). Wet soil coupled with overcast weather (low light intensity and poor air circulation) can cause an internal water pressure accumulation within leaves. Literally overnight, this water ruptures from the lower leaf surface causing scab-like spots (fig. 47). When the weather clears the condition stops, but damaged leaves will not recover.

Nutritional deficiencies are sometimes very similar to those caused by root pathogens such as fungi and nematodes (table 11). This is not hard to understand, since these diseases destroy portions of the root system, making it less capable of absorbing and transporting nutrients from the soil to the stems and leaves. The result of both conditions can be yellowing and

48 Fertilizer deficiency can make lower leaves of gloxina turn yellow.

49 Excess application of a foliar iron causes severe burning on this Gerbera daisy.

stunting (fig. 48). Soil and media pH may limit the availability of certain nutrients to plant roots even though they are present.

Alkaline soil or potting media (pH greater than 7.0) may cause minor nutrient deficiencies, such as iron chlorosis and manganese deficiency. Cold, wet soils in the spring may cause a temporary phosphorus deficiency that appears as leaf purpling. Nutritional excess or toxicity can appear similar to the effect of some leaf spotting fungi on plants that are fertilized regularly at relatively high rates (fig. 49).

Extremely low temperatures may damage soft foliage, quickly causing death of leaves and shoots. Rapid freezes may also cause root death on herbaceous plants that have not hardened off through a typical fall period. This damage may not be apparent until after the first growth flush the following spring.

High temperatures generated by changes in light levels may cause many plants to exhibit sunscald symptoms on tender leaves. This damage is short term until the plant acclimates to the new light level. On broad-leaved evergreens (such as rhododendron), marginal scorch commonly results from winter desiccation whether plants are in containers or in the ground.

Insects and mites can cause plant symptoms that are sometimes mistaken for disease. This is especially true of gall diseases (fig. 50). Bud mites that infest the growing tips of some annuals and perennials can cause distortion and stunting, which can be confused with herbicide injury or a virus (fig. 51). Unlike their larger relatives, the spider mites, these mites are too tiny to be seen with the naked eye, so diagnosis can be very difficult.

50 Spiney rose gall is induced by wasp larva.

51 Bud mites distort new growth on this *Ageratum*.

The appearance of mushrooms, slime mold, or birds' nest fungi is often correlated with the presence of decaying wood in a new mulch (fig. 52). These fungi are saprophytes feeding on the mulches and are not a threat (fig. 53). In a nursery such fungi may be unsightly and reduce product salability because customers think the plants may be damaged. Base control on reducing the amount of noncomposted wood and limiting wetness of potting media.

Many activities that are well-intentioned often have adverse effects on plants. Cleaning chemicals like muriatic acid (to clean brick and other masonry) may be toxic to nearby plants (fig. 54). Many lawn fertilizer and weed control products contain growth-regulator types of herbicides that

52 Slime mold can form on the soil, wood, or even plant leaves.

53 Bird's nest fungi on the soil surface.

54 Runoff of acid used to clean brick walls can create spots on nearby plants.

can drift by air movement or leach through the soil to damage nontargeted plants (fig. 55). Even our best efforts to control insects, mites, or diseases with legal and properly used pesticides may inadvertently damage nearby plants by spray drift. This damage is apt to develop within days of application and appear as a superficial spot burn on the outer leaves of a plant canopy (fig. 56).

Finally, a normal plant may exhibit symptoms that mimic those caused by a plant pathogen. Plants grown in milder climate zones may appear to be evergreen rather than deciduous as they would be in the northern United

55 Growth regulators can distort *Pelargonium* (geranium) leaves.

56 Herbicide phytotoxicity on *Hemerocallis* (daylily).

57 A natural sport or chimera on a portion of vinca.

States. All plant parts (roots, leaves, and flowers) are subject to the aging process (called senescence) regardless of where they are grown. However, herbaceous plants grown in Florida may retain their leaves longer than if they were grown in Michigan. Leaves age, become chlorotic, and die, and senescence-triggered leaf drop may resemble that caused by a root disease.

Another phenomenon of a normal plant is the natural genetic mutation that plants share with all living organisms. A genetic abnormality like variegation is quite common in some plants and can be confused with symptoms of virus diseases (fig. 57). Some of the most popular cultivars and selections are the result of these mutations. View these abnormalities as advantageous, not problematical.

Table 1

Scientific and common names of annual and perennial flowers

Scientific name	Common name
Abelmoschus, Malva	Musk mallow
Achillea	Yarrow
Agapanthus	African lily, Lily-of-the-Nile
Ageratum	Flossflower
Ajuga	Bugleweed
Alcea	Hollyhock
Amaryllis	Belladonna lily
Anemone	Lily-of-the-field, Windflower
Anigozanthos	Kangaroopaw
Antirrhinum	Snapdragon
Aquilegia	Columbine
Artemisia	Dusty miller, Wormwood
Aruncus	Goatsbeard
Asclepias	Butterfly weed, Milkweed
Aster	Frost flower, Michaelmas daisy, Starwort
Astilbe	False spiraea, Garden spirea
Barleria	Philippine violet
Begonia	Begonia
Bellis	English daisy
Bergenia	Bergenia
Bougainvillea	Bougainvillea
Buddleia	Butterfly bush
Caladium	Elephant's-ear
Calceolaria	Pocketbook plant, Slipperwort
Calendula	Pot marigold
Camellia	Camellia
Campanula	Bellflower
Canna	Garden canna
Capsicum	Pepper (ornamental)
Catharanthus	Madagascar periwinkle, Vinca
Celosia	Woolflower
Centaurea	Bachelor's button
Centranthus	Red valerian
Chrysanthemum maximum	Shasta daisy
Cimicifuga	Bugbane, Rattletop
Clematis	Clematis
Cleome	Spider flower
Coleus	Flame nettle, Painted leaves
Convallaria	Lily-of-the-valley
Coreopsis	Tickseed

Table 1. continued

Scientific name	Common name
Cortaderia	Pampas grass
Costus	Spiral flag
Crinum	Crinum lily, Spider lily
Crossandra	Crossandra
Dahlia	Dahlia
Delphinium	Larkspur
Dendranthema grandiflora (formerly Chrysanthemum x morifolium)	Florists' chrysanthemum
Dianthus	Carnation, Pink
Dicentra	Bleeding heart, Dutchman's-breeches
Dietes	African iris
Digitalis	Foxglove
Echinacea	Purple coneflower
Eucharis	Amazon lily
Euphorbia	Poinsettia
Eustoma	Lisianthus, Prairie gentian
Evolvulus	Blue daze
Forsythia	Golden-bells
Gaillardia	Blanket flower
Gardenia	Cape jasmine
Geranium	Cranesbill
Gerbera	Gerber daisy, Transvaal daisy
Geum	Avens
Gladiolus	Gladiolus
Gloriosa	Malabar glory lily
Gomphrena	Globe amaranth
Gypsophila	Babysbreath
Hedera	English ivy
Helianthus	Sunflower
Heliopsis	False sunflower, Oxeye
Hemerocallis	Daylily
Heuchera	Alumroot, Coral bells
Hibiscus	Hibiscus, Rose mallow, Swamp mallow
Hippeastrum	Amaryllis
Hosta	Funkia, Plantain lily
Hydrangea	Hydrangea
Impatiens	Garden balsam, Impatiens
Ipomoea	Morning glory, Railroad vine
Iris	Iris
Ixora	Chinese ixora
Jasminum	Jasmine
Kniphofia	Red-hot poker, Torch lily
Lavandula	Lavender
Leucothoe	Fetterbush
Liatris	Blazing-star, Gay-feather

continued

Table 1. continued

Scientific name	Common name
Lilium	Lily (tiger, trumpet etc.)
Limonium	Sea lavender, Statice
Liriope	Lilyturf
Lupinus	Lupine
Lychnis	Rose campion
Lysimachia	Circleflower
Lythrum	Loosestrife
Malva	Mallow
Mandevilla	Mandevilla
Matthiola	Stock
Monarda	Bee balm, Bergamot
Nasturium	Watercress
Ophiopogon	Border grass
Oxalis	Oxalis
Pachysandra	Japanese spurge
Paeonia	Peony
Pelargonium	Geranium
Penstemon	Beard-tongue
Pentas	Star-cluster
Petunia	Petunia
Phlox	Annual phlox, Phlox
Physostegia	False dragonhead
Pisum	Sweet pea
Platycodon	Balloon flower
Potentilla	Cinquefoil, Five-finger
Primula	Primrose
Ranunculus	Buttercup
Rhododendron	Azalea
Rosa	Rose
Rudbeckia	Black-eyed Susan, Coneflower
Salvia	Sage
Senecio	Mexican flame vine
Silene	Campion, Catchfly
Stokesia	Stokes' aster
Strelitzia	Bird-of-paradise
Tagetes	Marigold
Thunbergia	Clock vine
Trachelospermum	Asiatic jasmine
Tradescantia	Spiderwort
Tricyrtis	Toad lily
Trollius	Globeflower
Tulipa	Tulip
Verbena	Verbena
Veronica	Veronica

Table 1. continued

Scientific name	Common name
Vinca	Periwinkle
Viola	Pansy, Violet
Wedelia	Wedelia
Wisteria	Chinese wisteria
Yucca	Adam's needle, Spanish bayonet
	Spineless yucca
Zingiber	Ginger
Zinnia	Zinnia

Table 2

Cultural and other nonchemical control strategies for diseases of annuals and perennials, grouped by disease type

Disease type	Control strategies
Foliar diseases	
Fungal leaf spot or blight	Remove damaged leaves when the plants are not wet and destroy or discard them immediately. Keep plant leaves as dry as possible, and water early in the day to reduce the amount of time leaves are wet. Use drip or ground irrigation to reduce spread from splashing. Space plants to allow good air circulation.
Bacterial leaf spot or blight	Remove damaged leaves when the plants are not wet and destroy or discard them immediately. Keep plant leaves as dry as possible, and water early in the day to reduce the amount of time the leaves are wet. Use drip or ground irrigation to reduce spread from splashing. Space plants to allow good air circulation.
Rust	Remove damaged leaves when the plants are not wet and destroy or discard them immediately. Avoid crowding plants and space them to allow good air circulation. Keep plant leaves as dry as possible, and water early in the day to reduce the amount of time the leaves are wet. Use resistant plant varieties if they are known.
Powdery mildew	Avoid crowding plants and space them to allow good air circulation. Keep plant leaves as dry as possible, and water early in the day to reduce the amount of time the leaves are wet. Grow plants in full sun if possible. Use resistant plant varieties if they are known.
Downy mildew	Avoid crowding plants and space them to allow good air circulation. Keep plant leaves as dry as possible, and water early in the day to reduce the amount of time the leaves are wet. Grow plants in full sun if possible. *continued*

Table 2. continued

Disease type	Control strategies

Root and stem diseases

Root and stem rot	Remove and destroy or discard the entire plant including the soil immediately surrounding its roots. For future ground plantings, improve soil drainage or use a looser potting media using new soil or media if possible. Check that the roots and stems of new plants are free of symptoms.
Damping-off	Use clean, new pots or flats and a potting media with good drainage. Do not stress seedlings with too much or too little water. Discard pots or flats with plants showing symptoms of damping-off. Plant seed at appropriate potting medium or soil temperatures to maximize seed germination percentage and speed.
Crown gall	Remove and destroy or discard the entire plant including the soil immediately surrounding its roots. Replant with a species that is less susceptible to crown gall, if possible. Use new soil or potting medium. Check roots and stems of new plants to make sure they are free of symptoms. Never propagate from plants with symptoms of crown gall.
Canker or dieback	Prune away stems with canker or dieback symptoms and discard or destroy. Clean pruners between cuts with rubbing (isopropyl) alcohol. Never cut through a canker but well below it. Do not apply pruning paint to cut ends.
Vascular wilt	Remove and destroy or discard the entire plant including the soil immediately surrounding its roots. Replant ground beds with a non-susceptible species, if possible. Improve soil drainage or use a looser potting medium, using new soil or potting medium if possible. Never use cuttings from plants with symptoms of vascular wilt. In warmer climates, soil solarization can be helpful. Cover beds with clear plastic mulch to allow heat buildup for as long as feasible.

Virus and mycoplasma diseases

Virus	Remove and destroy symptomatic plants. Keep insect and mite pests under control. Remove weeds. Never use cuttings from plants with symptoms of virus infection.
Aster yellows	Remove and destroy symptomatic plants. Keep leafhoppers under control and remove weeds. Never use cuttings from plants with symptoms of aster yellows.
Nematode	Remove and destroy or discard the entire plant including the soil immediately surrounding its roots. Use new soil or potting medium if possible. Check that the roots and stems of new plants are free of symptoms. Plant resistant cultivars when possible. For foliar nematode infestations, remove infected leaves or the entire plant to slow spread to other plants.

Table 3

Annuals and perennials susceptible to *Corynespora cassiicola*

Scientific name	Common name
Ajuga reptans	Bugleweed
Allamanda spp.	Allamanda
Antirrhinum majus	Snapdragon
Begonia spp.	Begonia
Capsicum spp.	Pepper (ornamental)
Catharanthus roseus	Madagascar periwinkle
Coleus x *hybridus*	Coleus
Crossandra spp.	Crossandra
Digitalis spp.	Foxglove
Euphorbia pulcherrima	Poinsettia
Evolvulus glomeratus	Blue daze
Hydrangea spp.	Hydrangea
Impatiens spp.	Impatiens
Jasminus spp.	Jasmine
Justicia spp.	Shrimp plant
Lantana spp.	Lantana
Liatris sp.	Gay-feather
Mandevilla spp.	Mandevilla
Monarda punctata	Beebalm
Ocimum basilicum	Basil
Pachystachys spp.	Cardinal's guard
Petunia x *hybrida*	Petunia
Rhododendron spp.	Azalea
Salvia spp.	Sage
Sedum spp.	Sedum
Strelitzia spp.	Bird-of-paradise
Trachelospermum jasminoides	Asiatic jasmine
Wedelia trilobata	Wedelia
Wisteria sinensis	Chinese wisteria
Zantedeschia sp.	Calla lily

Table 4

Annuals and perennials susceptible to *Myrothecium roridum*

Scientific name	Common name
Ajuga reptans	Bugleweed
Anthurium spp.	Tailflower
Antirrhinum majus	Snapdragon
Begonia spp.	Begonia
Bougainvillea spp.	Bougainvillea
Delphinium spp.	Larkspur
Dendranthema grandiflora (formerly *Chrysanthemum* x *morifolium*)	Florists' chrysanthemum
Echinacea purpurea	Purple coneflower
Euphorbia pulcherrima	Poinsettia
Gardenia jasminoides	Cape jasmine
Gerbera spp.	Transvaal daisy
Impatiens spp.	Impatiens
Jasminium spp.	Jasmine
Lantana camara	Lantana
Matthiola spp.	Stock
Mentha spp.	Mint
Ocimum basilicum	Basil
Passiflora spp.	Passion vine
Petunia x *hybrida*	Petunia
Rudbeckia spp.	Coneflower
Sedum spp.	Sedum
Selaginella spp.	Little club moss
Streptocarpus spp.	Cape primrose
Verbena x *hybrida*	Verbena
Zingiber darceyi	Ginger (variegated)

Table 5

Annuals and perennials susceptible to *Erwinia* species

Scientific name	Common name
Agapanthus spp.	African lily
Agave spp.	Century plant
Begonia spp.	Begonia
Belamcanda chinensis	Blackberry lily

Table 5. continued

Scientific name	Common name
Caladium spp.	Elephant's ear
Capsicum spp.	Pepper (ornamental)
Commelina spp.	Dayflower
Dahlia pinnata	Dahlia
Dendranthema grandiflora (formerly *Chrysanthemum* x *morifolium*)	Florists' chrysanthemum
Dianthus spp.	Carnation
Eucharis grandiflora	Amazon lily
Euphorbia pulcherrima	Poinsettia
Gerbera spp.	Transvaal daisy
Gladiolus x *hortulanus*	Gladiolus
Gomphrena globosa	Globe amaranth
Hedera helix	English ivy
Helianthus spp.	Sunflower
Hemerocallis spp.	Daylily
Hibiscus spp.	Swamp mallow
Hosta spp.	Plantain lily
Hydrangea spp.	Hydrangea
Impatiens spp.	Impatiens
Iris x *germanica*	Iris
Lilium lancifolium	Lily (tiger)
Lilium longiflorum	Lily (trumpet)
Limonium spp.	Statice
Mentha spp.	Mint
Mesembryanthemum crystallinum	Ice plant
Narcissus spp.	Narcissus
Pelargonium x *hortorum*	Zonal geranium
Petunia x *hybrida*	Petunia
Plumbago auriculata	Cape leadwort
Polianthes tuberosa	Tuberose
Portulaca grandiflora	Rose moss
Portulaca oleracea	Purslane
Primula spp.	Primrose
Rhoeo spathacea	Oyster plant
Ruscus spp.	Butcher's broom
Sedum spp.	Sedum
Senecio sp.	Mexican flame vine
Strelitzia reginae	Bird-of-paradise
Tagetes spp.	Marigold
Tradescantia spp.	Spiderwort
Yucca spp.	Adam's needle
Zantedeschia spp.	Calla lily
Zebrina spp.	Wandering Jew
Zinnia elegans	Zinnia

Table 6

Annuals and perennials susceptible to *Pseudomonas cichorii*

Scientific name	Common name
Anthurium sp.	Tailflower
Argemone mexicana	Mexican poppy
Barleria cristata	Phillipine violet
Caladium spp.	Elephant's ear
Calendula officinalis	Pot marigold
Capsicum spp.	Pepper (ornamental)
Chrysanthemum maximum	Shasta daisy
Coreopsis spp.	Tickseed
Delphinium spp.	Larkspur
Dendranthema grandiflora (formerly *Chrysanthemum* x *morifolium*)	Florists' chrysanthemum
Euphorbia pulcherrima	Poinsettia
Gaillardia pulchella	Blanket flower
Gazania spp.	Treasure flower
Geranium spp.	Cranesbill
Gerbera spp.	Transvaal daisy
Hedera helix	English ivy
Hibiscus spp.	Swamp mallow
Hydrangea spp.	Hydrangea
Impatiens spp.	Impatiens
Jasminum spp.	Jasmine
Justicia spp.	Shrimp plant
Mentha spp.	Mint
Ocimum basilicum	Basil
Pachystachys lutea	Cardinal's guard
Pelargonium x *hortorum*	Zonal geranium
Petunia x *hybrida*	Petunia
Rhododendron spp.	Azalea
Rosa spp.	Rose
Ruscus spp.	Butcher's broom
Salvia spp.	Sage
Tagetes spp.	Marigold

Table 7

Annuals and perennials susceptible to *Pseudomonas syringae* pv. *syringae*

Scientific name	Common name
Abelmoschus esculentus	Musk mallow
Capsicum spp.	Pepper (ornamental)
Catharanthus roseus	Madagascar periwinkle
Chrysopsis spp.	Golden aster
Dendranthema grandiflora (formerly *Chrysanthemum* x *morifolium*)	Florists' chrysanthemum
Hedera helix	English ivy
Hibiscus spp.	Swamp mallow
Impatiens spp.	Impatiens
Liatris spicata	Gay-feather
Pelargonium x *hortorum*	Zonal geranium
Ranunculus spp.	Buttercup
Rosa spp.	Rose
Salvia splendens	Scarlet salvia
Strelitzia reginae	Bird-of-paradise
Tagetes spp.	Marigold

Table 8

Annuals and perennials susceptible to *Sclerotium rolfsii*

Scientific name	Common name
Abelmoschus esculentus	Musk mallow
Ageratum spp.	Flossflower
Ajuga spp.	Bugleweed
Alcea rosea	Hollyhock
Alternanthera spp.	Joseph's coat
Amaryllis belladonna	Belladonna lily
Anemone spp.	Windflower
Antirrhinum majus	Snapdragon
Aquilegia sp.	Columbine
Arctotis stoechadifolia	African daisy
Armeria spp.	Plumbago
Barleria spp.	Phillipine violet
Begonia x *semperflorens-cultorum*	Begonia
Buddleia spp.	Butterfly bush
Caladium spp.	Elephant's ear
Calendula officinalis	Pot marigold
Campanula divaricata	Bellflower
Canna sp.	Garden canna
Capsicum spp.	Pepper (ornamental)
Centaurea cyanus	Bachelor's button
Coreopsis spp.	Tickseed
Cosmos spp.	Cosmos
Dahlia pinnata	Dahlia
Delphinium spp.	Larkspur
Dendranthema grandiflora (formerly *Chrysanthemum* x *morifolium*)	Florists' chrysanthemum
Dianthus barbatus	Sweet William
Dianthus caryophyllus	Carnation
Dicentra sp.	Bleeding heart
Digitalis spp.	Foxglove
Dimorphotheca sp.	Cape marigold
Echinops sp.	Thistle
Erigeron sp.	Fleabane
Eucharis grandiflora	Amazon lily
Euphorbia pulcherrima	Poinsettia
Fescue spp.	Fescue
Gardenia jasminoides	Cape jasmine
Gerbera jamesonii	Transvaal daisy
Gilia spp.	Gilia
Gladiolus x *hortulanus*	Gladiolus

Table 8. continued

Scientific name	Common name
Gloriosa rothschildiana	Malabar glory lily
Hedera helix	English ivy
Helianthus spp.	Sunflower
Helichrysum bracteatum	Everlasting
Hemerocallis spp.	Daylily
Hibiscus spp.	Swamp mallow
Hippeastrum spp.	Amaryllis
Hosta spp.	Plantain lily
Hydrangea spp.	Hydrangea
Iberis sp.	Candytuft
Impatiens spp.	Impatiens
Jasminum spp.	Jasmine
Kniphofia spp.	Red hot poker
Lantana involucrata	Lantana
Lathyrus odoratus	Sweet pea
Lavatera arborea	Tree mallow
Liatris spp.	Gay-feather
Lilium spp.	Lily
Limonium spp.	Statice
Linaria canadensis	Toadflax
Linum sp.	Flax
Liriope muscari	Lilyturf
Lupinus spp.	Lupine
Lysimachia spp.	Circleflower
Matthiola sp.	Stock
Monarda sp.	Bee balm
Narcissus spp.	Narcissus
Nepeta sp.	Catnip
Nymphaea dorata	Waterlily
Ocimum basilicum	Basil
Ophiopogon japonicus	Bordergrass
Oxalis spp.	Wood sorrel
Paeonia sp.	Peony
Passiflora spp.	Passion vine
Penstemon sp.	Beard-tongue
Pentas spp.	Star-cluster
Petunia x *hybrida*	Petunia
Phlox spp.	Phlox
Physalis sp.	Chinese lantern
Physostegia spp.	False dragonhead
Polianthes tuberosa	Tuberose
Ranunculus spp.	Buttercup
Rudbeckia spp.	Coneflower
Salvia spp.	Sage
Scabiosa atropurpurea	Pincushion flower

continued

Table 8. continued

Scientific name	Common name
Senecio spp.	Mexican flame vine
Silene sp.	Campion
Stokesia laevis	Stoke's aster
Tagetes spp.	Marigold
Thunbergia spp.	Clock vine
Tithonia rotundifolia	Mexican sunflower
Tulbaghia violacea	Lily
Tulipa gesnerana	Tulip
Veronica spp.	Veronica
Vinca spp.	Periwinkle
Viola spp.	Violet
Viola tricolor	Pansy
Xantheranthemum igneum	Immortelle
Yucca spp.	Adam's needle
Zantedeschia aethiopica	Calla lily
Zephyranthes spp.	Rain lily
Zingiber officinale	Ginger
Zinnia elegans	Zinnia

Table 9

Annuals and perennials susceptible to *Agrobacterium tumefaciens*

Scientific name	Common name
Begonia spp.	Begonia
Bougainvillea spp.	Bougainvillea
Camellia japonica	Camellia
Coleus x *hybridus*	Flame nettle
Dahlia spp.	Dahlia
Dendranthema grandiflora (formerly *Chrysanthemum* x *morifolium*)	Florists' chrysanthemum
Gypsophila spp.	Babysbreath
Hibiscus spp.	Swamp mallow
Jasminum spp.	Jasmine
Lantana camara	Lantana
Malva spp.	Mallow
Pachystachys lutea	Cardinal's guard
Pentas spp.	Star-cluster
Phlox drummondii	Annual phlox
Plumbago spp.	Leadwort
Rosa spp.	Rose
Thunbergia spp.	Clock vine
Wisteria sinensis	Chinese wisteria

Table 10

Annuals and perennials susceptible to Cucumber mosaic virus

Scientific name	Common name
Agave spp.	Century plant
Allamanda cathartica	Allamanda
Amaryllis belladonna	Belladonna lily
Antirrhinum majus	Snapdragon
Capsicum spp.	Pepper (ornamental)
Catharanthus roseus	Madagascar periwinkle
Celosia argentea	Woolflower
Cheiranthus cheiri	Wallflower
Commelina spp.	Dayflower
Dendranthema grandiflora (formerly *Chrysanthemum morifolium*)	Florists' chrysanthemum
Gladiolus x *hortulanus*	Gladiolus
Gloriosa spp.	Malabar glory lily
Gomphrena globosa	Globe amaranth
Helianthus annuus	Sunflower
Hippeastrum spp.	Amaryllis
Iris x *germanica*	Iris
Lilium longiflorum	Lily (trumpet)
Lupinus spp.	Lupine
Matthiola incana	Stock
Passiflora edulis	Passion vine
Pelargonium peltatum	Ivy geranium
Pelargonium x *hortorum*	Zonal geranium
Petunia x *hybrida*	Petunia
Portulaca oleracea	Purslane
Scaevola spp.	Beach Naupaka
Tagetes spp.	Marigold
Viola odorata	Violet

Table 11

Symptoms of nutrient deficiency

Nutrient	Chemical symbol	Symptoms
Nitrogen	N	Plants are pale green or yellow; small leaves; oldest leaves yellow first; stunting or poor growth in general.
Phosphorus	P	Plant is dark green with a bronzing or purple tinge; leaves small; stunting.
Potassium	K	Older leaves are mottled often with a yellow or brown leaf margin; leaf and stem size reduced.
Magnesium	Mg	Lower leaves have interveinal and marginal yellowing and browning.
Calcium	Ca	Dead tips and margins are on young leaves; leaves may be hooked at tips.
Iron	Fe	Young leaves show interveinal yellowing; larger leaf veins remain green.
Sulfur	S	Leaves are slightly green with veins lighter than surrounding tissue; weak growth on some plants.
Manganese	Mn	Dead spots are scattered across leaves; yellowing of new leaves except for small veins gives leaves a netted appearance; color and size of flowers poor.
Boron	B	Terminal leaves are distorted, brittle, and small.
Copper	Cu	New growth is severely distorted; terminal buds die and sometimes a witches'-broom forms.

ACHILLEA (YARROW)

58 Dark brown rust pustules form anywhere on yarrow leaves.

59 Powdery mildew appears like white frosty patches of fungal growth on Angel's breath yarrow (*Achillea ptarmica*).

AGAPANTHUS

60 Phytophthora stem rot turns *Agapanthus* basal leaves yellow and brown and makes the stems rot.

Agapanthus

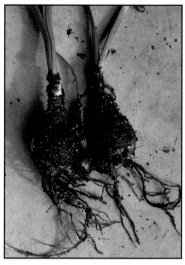

61 Virus-like symptoms on *Agapanthus* are the light green or white areas.

62 *Pythium* causes roots to become blackened and mushy.

Ageratum

63 Botrytis blight turns flowers brown. They also become covered with the grayish spores of the fungus.

Ageratum

65 *Pythium* and *Fusarium* rot plant roots causing leaves to wilt and develop tip burn.

64 *Fusarium* and *Rhizoctonia* can produce brown, sunken spots at *Ageratum* stem bases.

Alcea (hollyhock)

66 Cercosporella leaf spot is characterized by gray to black spots that join and ◄ cause leaves to tatter.

ALCEA (HOLLYHOCK)

67 Hollyhock upper leaf surfaces show yellow rust pustules, while lower leaf surfaces show gray pustules.

68 In severe rust infections, leaves become blighted.

ANEMONE

69 Black sclerotia form in rotted stem tissue on anemone infected with *Sclerotinia*.

70 Light green color breaks and ringspots are typical virus symptoms on Japanese anemone.

ANTIRRHINUM (SNAPDRAGON)

71 Grayish patches on upper and lower leaf surfaces are signs of downy mildew.

72 Yellow blotches on upper leaf surfaces signify the presence of rust pustules on undersides.

73 A close-up of dark brown, dusty rust pustules on snapdragon.

74 One-sided death is a common sign of a vascular wilt, such as Verticillium wilt on snapdragon.

75 Colletotrichum or frog-eye spot on snapdragon is typified by tan spots with a distinct light eye that is almost one-half inch in diameter.

ANTIRRHINUM (SNAPDRAGON)

76 Damping-off causes small seedlings to ◄ rot at the soil line.

77 Cercospora leaf spot on snapdragon shows up as small tan spots with purple margins. ➤

78 Impatiens necrotic spot virus appears as round necrotic ringspots ◄ on leaves.

AQUILEGIA (COLUMBINE)

79 Ascochyta leaf spot on *Aquilegia* causes dark areas with
◄ purple margins.

80 Powdery mildew fungus appears like white frosty patches that can cover entire
◄ leaflets.

81 Powdery mildew sometimes turns infected leaves purplish.

AQUILEGIA (COLUMBINE)

82 Leaf miner injury on *Aquilegia* appears as white squiggly trails with a papery texture.

83 Spider mite feeding causes light speckled areas on the topsides of leaves.

84 Leaves infected with foliar nematode have dark brown, vein-delimited areas.

Aquilegia (columbine)

85 *Botrytis* and *Alternaria* can cause tan spots with darker margins to form on leaves; sometimes entire leaflets
◄ will turn yellow.

Aster

86 Rosy blister gall on wild aster is caused by an insect (*Asteromyzia rosea*).

87 The thin, yellow stem with small, white flower cluster is dodder, a parasitic plant.

88 On some plants powdery mildew produces purplish discoloration and leaf
◄ distortion.

89 Orange rust pustules on leaf
◄ undersides.

90 Anthracnose makes leaf spots that are irregularly shaped and dark
▼ brown.

ASTER

91 Sulfur dioxide (air pollution) damage causes the upper leaf epidermis to turn white and papery.

92 Aster plants infected with aster yellows mycoplasma are stunted and have yellow foliage (plant on the left is healthy).

93 Dry, brown stem cankers from *Phomopsis* form on bases of infected aster.

94 *Rhizoctonia* forms brown cankers at stem bases.

ASTILBE

95 Tarnished plant bug damage on *Astilbe*.

96 Large areas of plant foliage die when *Astilbe* are subjected to drought.

97 Both *Pseudomonas* and *Xanthomonas* cause brown, irregularly shaped spots to form on leaf margins and centers.

98 Lacebug damage on azalea appears as speckling on upper leaf ◄ surface.

99 Adult lace-bugs can sometimes be found on leaf ◄ undersides.

100 Cercospora leaf spot appears as reddish-brown spots of different ◄ sizes.

101 Taxus weevils damage azalea stems by girdling them near the soil line.

102 Scorch develops on leaves when plants become desiccated.

103 Gray, irregularly shaped galls form on azalea stem tips infected with leaf gall fungus (*Exobasidium vaccini*).

104 Immature galls of the leaf gall fungus are light green and thickened.

105 Copper deficiency on azalea creates tiny, yellow, new leaves. ◀

106 Azalea iron deficiency causes stunted, very yellow plants. ▶

107 Brown, dead areas between veins form on azalea leaves infected with foliar nematode. ◀

AZALEA

108 Cold injury causes corky galls to form on stems.

109 Ovulinia petal blight on azalea first appears as clear pale spots on petals.

110 Extensive flower drop can result from severe petal blight infections.

111 Botrytis flower infections can extend into the stems and cause tip ◄ dieback.

112 Entire plants may become covered with white fungal growth when infected with powdery mildew. ➤

113 Sunken brown areas form at the soil line of cuttings ◄ infected with *Rhizoctonia*.

BEGONIA

114 Frosty white patches of powdery mildew fungus form mainly on upper leaf surfaces. ➤

115 Impatiens necrotic spot virus causes begonia leaves to become mottled and develop dead ◄ spots.

116 Chlorotic mottling can also result from impatiens necrotic spot virus. ➤

117 Cuttings can rot from root or stem rot moving upwards.

118 Large quantities of white cottony mycelia cover begonias infected with *Sclerotinia*.

119 Botrytis blight appears as clear sunken areas on petals.

BEGONIA

120 Leaf spots from *Botrytis cinerea* are irregularly shaped and scattered across the leaf.

121 Bacterial leaf spot on this tuberous begonia appears as roughly circular tan spots with darker margins on all age leaves.

122 Bacterial leaf spot infection on flower petals is characterized by dark
◄   brown areas.

BERGENIA

123 Leaf bases of *Bergenia* infected with anthracnose become distorted and die. ◄

124 Close-up of taxus weevil damage on leaf edges. ◄

125 Ringspots and purple mottling form primarily on *Bergenia* leaf edges infected with an unidentified virus.

126 Tiny chlorotic areas are the first symptoms of Alternaria leaf spot on *Bergenia*.

BERGENIA

127 *Bergenia* leaves have large, brown, vein-delimited areas when infected with foliar
◄ nematode.

BOUGAINVILLEA

128 Pseudocercospora leaf spots on bougainvillea are tan with reddish margins. ➤

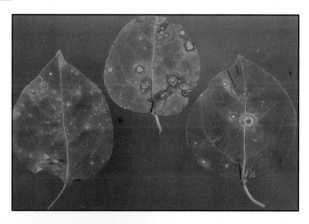

129 Pseudomonas leaf spots are reddish-brown with yellow halos and form on leaf margins and
◄ centers.

BOUGAINVILLEA

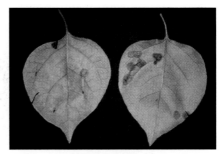

130 Cucumber mosaic virus causes leaves to become distorted and mottled.

131 Rhizoctonia leaf spots are round, gray and up to one-half inch in diameter.

CALENDULA

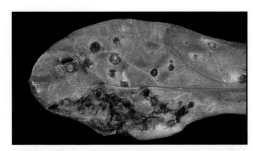

132 Calendula smut is typified by rounded, ◄ brown leaf spots.

133 Alternaria leaf spot on calendula appears as reddish-purple spots that can reach one-fourth inch wide.

CALENDULA

134 Calendula infected with aster yellows mycoplasma have green flowers and are stunted.

135 Frosty patches of white powdery mildew can turn plants white. ➤

CAMELLIA

136 Large brown areas appear on upper portions of exposed leaves when camellia ◄ are sunburned.

138 Grayish-green circular patches on upper leaf surfaces (especially in warm, wet climates) are signs of algal leaf spot.

137 Advanced anthracnose infections on camellia can result in stem dieback.

139 Camellia variegation virus commonly appears with bright yellow or white markings on leaves. ➤

140 Ringspots on camellia can be green on a ◄ yellow leaf.

141 Exobasidium fungus causes leaves to become swollen, sometimes pinkish and distorted.

142 Root rot makes camellias appear stunted and unthrifty and can even kill them.

143 Colletotrichum leaf spot starts on wounded areas or at leaf margins and can have concentric rings of light and dark tissue.

◀

144 Sooty mold on camellia is produced when a saprophytic fungus grows on honeydew from an insect infestation.

CAMELLIA

145 Lower leaf surfaces of camellia infested with tea scale are white, while upper leaf surfaces are yellow.

146 Another scale shows individual scale insects as raised areas clustered along leaf veins.

CANNA

147 Canna rust pustules are orange and form on both leaf surfaces.

148 Xanthomonas leaf spot starts at leaf margins and can spread along veins into the centers.

CANNA

149 Xanthomonas leaf spot on canna can sometimes appear as discrete brown spots surrounded by a yellow halo.

150 Bean yellow mosaic virus on infected cannas causes mosaic and streaking.

CATHARANTHUS (VINCA)

151 Impatiens necrotic spot virus on vinca severely distorts and blackens new leaves.

CATHARANTHUS (VINCA)

152 Dieback in the landscape is caused by a variety of soilborne fungi. ◄

153 Vinca can become stunted and yellowed with severely distorted tips from aster yellows mycoplasma. ➤

154 Alternaria leaf spot can appear as dark brown or black spots on leaves, stems, and flowers.

CATHARANTHUS (VINCA)

155 In severe infections, vinca with Alternaria leaf spot ◄ lose their lower leaves.

156 *Rhizopus* on vinca causes severe blighting during the warmer seasons and can be identified by the matted, web-like growth covering the plants.

157 Sclerotia (black structure in the center) form in stem tissues of vinca infected with ◄ *Sclerotinia.*

CATHARANTHUS (VINCA)

158 Phytophthora aerial blight in the landscape can result in a complete loss of a vinca planting.

159 Phytophthora aerial blight on potted vinca causes wilt and collapse of infected tissues.

CELOSIA

160 Pythium root rot on celosia creates stunting, yellowing, and wilting of leaves and flowers.

161 Botrytis blight of celosia flowers displays brown ◄ sunken areas.

162 Fusarium can cause cankers at celosia stem base. ➤

163 Alternaria leaf spot appears on leaf margins as dark brown areas with concentric rings of light and dark tissue. ◄

164 *Corynebacterium fascians* causes Shasta daisy shoots to proliferate and never grow properly. ➤

165 Stem and root infections of Shasta daisy in pots result in death of many plants. ◄

CHRYSANTHEMUM (SHASTA DAISY)

166 Septoria leaf spot on Shasta daisy forms
◄ mainly on leaf tips and margins.

167 *Erwinia* causes black, mushy spots over Shasta daisy leaves.

168 Tips of Shasta daisy leaves are infected with *Botrytis*.

CHRYSANTHEMUM (SHASTA DAISY)

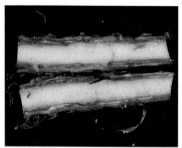

169 Cross sections of Shasta daisy stems infected with Acremonium wilt show brown discolored streaks in the vascular tissue.

170 Acremonium wilt reveals typical symptoms of a vascular disease.

CHRYSANTHEMUM (FLORIST'S MUM)

171 Overfertilized plants (front) can appear yellow and stunted.

CHRYSANTHEMUM (FLORIST'S MUM)

172 Tomato spotted wilt virus on chrysanthemum causes new leaves to be stunted with black spots.

173 Botrytis blight of stored mums can develop from a single infection point and form masses of grayish mycelia and spores.

174 White rust on mum results from ◄ *Puccinia horiana.*

175 Boron deficiency (right and center) causes mum leaves to be stunted and form yellow and brown margins. The leaf on the left is healthy. ➤

CHRYSANTHEMUM (FLORIST'S MUM)

176 Brown rust on chrysanthemum creates dark reddish-brown pustules primarily on leaf undersides.

177 Fusarium wilt on mum shows death of leaves on one side of plant.

178 Wedge-shaped spots are caused ◀ by foliar nematode infections.

179 Foliar nematode can produce leaf distortion. ➤

CHRYSANTHEMUM (FLORIST'S MUM)

180 Thrips feeding injury appears as etching on upper leaf surfaces.

181 Ascochyta petal blight can result in dark brown petal tips.

182 Pseudomonas leaf spot creates black watery spots on plants.

183 Both *Rhizoctonia* and *Pythium* can cause cutting bases to rot and turn mushy.

CHRYSANTHEMUM (FLORIST'S MUM)

184 *Fusarium* can also attack rooted plants and produce a stem canker at the soil line.

CHRYSANTHEMUM SP.

185 *Chrysanthemum pacificum* infected with *Rhizoctonia solani* shows gradual loss of stems through basal rot.

186 *Ascochyta* causes discrete, black, irregularly shaped spots
◄ on clematis leaves.

187 Wilt can be an early sign of *Ascochyta* infection on clematis.

188 Cyclamen mite infestation shows stunting and severe distortion of
◄ new shoot tissues.

COREOPSIS

190 Cercospora leaf spot on coreopsis is usually found on lower leaves.

189 Frosty patches of powdery mildew form on upper
◄ leaf surfaces of this coreopsis.

191 This coreopsis has
a mixed infection of
Alternaria and
Rhizoctonia causing
◄ leaf spots.

192 *Pseudomonas cichorii*
infections are very common
on coreopsis and typically
appear as greasy gray or
black spots on lower leaves
that can spread to kill entire
leaves under warm, wet
conditions. ➤

COREOPSIS

193 Aster yellows disease on a Sunray coreopsis causes the virescent (green) flowers on plants at right.

COSMOS

194 *Botrytis* causes many small brown spots on cosmos petals. ◄

195 The cosmos on the left shows stunting and yellowing, typical signs of aster yellows. ➤

196 *Xanthomonas* infections on *Crossandra* appear mainly on leaf margins. ➤

197 *Rhizoctonia* spp. kill roots of many plants, turning them black ◄ and mushy.

198 *Botrytis* can produce blight on *Crossandra* leaves and buds. ➤

CROSSANDRA

199 The bright yellow mosaic and ringspot patterns on this
Crossandra are probably caused by a virus.

DAHLIA

200 Impatiens necrotic spot virus on dahlia appears
as light green mottling.

201 *Alternaria* causes
discrete gray to black
spots with concentric
rings of light and dark
tissue.

202 The speckled white appearance of this
leaf indicates spider mite infestation on
◄   leaf undersides.

203 In severe cases, spider mites form webs across leaves and stems. Tiny yellowish specks are spider mites. ◄

204 Yellow mottling is caused by planthopper feeding and not a virus. ➤

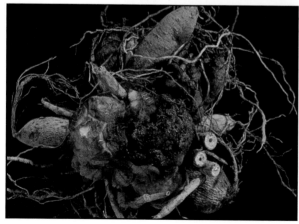

205 Crown gall bacterium on dahlia produces galls at stem bases. ◄

DAHLIA

206 Smut causes tan spots with darker borders to form on Rigoletto dahlia. ➤

207 Powdery mildew on ◄ dahlia.

208 Tomato spotted wilt virus appears as distinct chlorotic mottling on part of dahlia plant. ➤

DELPHINIUM (LARKSPUR)

209 Cyclamen mite infestation blackens *Delphinium* buds. ◄

210 *Sclerotium rolfsii* kills *Delphiniums*. Note the sclerotia (round, seedlike structures) at base of infected stem. ➤

211 *Ascochyta* creates spots on *Delphinium* leaves and sometimes flowers. ◄

DELPHINIUM (LARKSPUR)

212 Powdery mildew infection of *Delphinium*. ◄

213 The small, circular, gray to brown spots come from *Pseudomonas cichorii*. ▼

DIANTHUS (CARNATION)

214 Many pathogens can cause root rot and death of *Dianthus* in the landscape. ◄

Dianthus (Carnation)

215 Alternaria leaf spot on Rainbow Lane *Dianthus* are small and surrounded by a reddish margin. ◄

216 Rust appears on leaf surfaces and stems as brown pustules that rupture the epidermis. ►

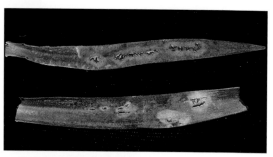

217 *Fusarium oxysporum* can cause wilt and death of *Dianthus* during production. ◄

DIANTHUS (CARNATION)

218 Rhizoctonia crown rot on *Dianthus* frequently kills the plant.

219 Advanced symptoms of Pythium root rot on
◀ *Dianthus*.

220 Fusarium wilt on *Dianthus* is typified by wilt and death of one section of the plant foliage. ▶

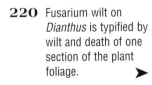

DICENTRA (BLEEDING HEART)

221 An unidentified virus in *Dicentra* causes a pale green mosaic ◄ pattern.

222 In ground beds, Verticillium wilt kills many ◄ *Dicentra*.

223 A close-up of Verticillium wilt symptoms on *Dicentra*. ➤

Echinacea (Purple Coneflower)

224 Angular brown spots on *Echinacea* are typical symptoms of bacterial leaf spot caused by *Pseudomonas cichorii.*

225 *Xanthomonas campestris* also produces a bacterial leaf spot on *Echinacea.*

226 Botrytis blight on *Echinacea*
◄ distorts young leaves.

227 During summer months, *Rhizopus* can infect flowers, giving them a grayish, fuzzy appearance. ➤

Echinacea (Purple Coneflower)

228 The yellow plant in the center did not receive ◄ any fertilizer.

Eustoma (Lisianthus)

229 *Botrytis* can cause brown spots in shoot ◄ tips of *Eustoma*.

230 *Botrytis* infections on *Eustoma* flowers produce water-soaked areas where tissue collapses.

231 Petal blight can be caused by *Colletotrichum*.

FORSYTHIA

232 Phytophthora canker on *Forsythia* creates brown to black areas that form ◄ near the ground.

233 *Sclerotinia* causes brown stem cankers and wilting. ►

234 Tarnished plant bug damage appears as black circular spots with holes in ◄ their centers.

235 Sunburn creates bleached areas on exposed fuchsia leaves. ➤

236 Botrytis leaf spots on fuchsia are tan with ◀ reddish borders.

237 Rust most often appears as orange pustules on leaf undersides. ➤

GAILLARDIA

238 Smut spots on *Gaillardia* are usually round, sunken, and tan or black. ◄

239 Septoria spots are white with tiny black specks that are the fruiting bodies of the fungus. ➤

240 Aster yellows mycoplasma causes *Gaillardia* flowers to become green and distorted. ◄

241 Iron deficiency on gardenia makes new leaves turn bright yellow.

242 *Phyllosticta* and *Myrothecium* cause roughly circular, brown dry spots on gardenia leaves and stems. ◄

243 A close view of Myrothecium leaf spot shows the black fungal fruiting bodies on the undersides of the spots. ➤

GARDENIA

244 Phytophthora root and stem rot results in loss of many gardenia cuttings.

245 When temperatures are high, lack of water can cause
◄ gardenia bud drop.

246 High soil pH can create minor element deficiencies in gardenia.

GERBERA (GERBER DAISY)

247 Broad mites and other tiny bud mites infesting *Gerbera* leaves can turn them ◄ brittle and stunted.

248 Botrytis blight on *Gerbera* flowers may affect both petals and flower centers. ➤

249 Powdery mildew can completely coat leaves.

250 Powdery mildew can create small grayish spots on *Gerbera* flowers.

GERBERA (GERBER DAISY)

251 *Pseudomonas cichorii* on leaves produces small black or tan spots that can enlarge and spread along leaf veins. ◄

252 Fusarium root and crown rot in a garden can kill *Gerbera*.

253 Rhizopus blight results in flowers covered with mycelia and spores.

GERBERA (GERBER DAISY)

254 Root rot turns *Gerbera* foliage yellowish.

255 Iron toxicity appears as marginal burning and stunting of new leaves.

256 Iron deficiency causes leaves to develop interveinal chlorosis.

Gerbera (Gerber Daisy)

258 *Phyllosticta* causes large, irregularly shaped tan spots with purplish borders.

257 Leaves infested with spider mites develop a speckled appearance, and the yellowish mites are found on leaf undersides near the veins.

Geum

260 Close-up of downy mildew on *Geum*.

259 Downy mildew creates angular yellow patches that appear on leaf uppersides.

GLADIOLUS

261 Cucumber mosaic virus produces white streaking and mosaic patterns on gladiolus flowers. ◄

262 Thrips damage on gladiolus leaves appears as white or tan etched areas. ➤

263 Gladiolus with *Xanthomonas* infections develop brown to black angular spots with yellow margins on leaves. ◄

GOMPHRENA

264 *Xanthomonas* causes small reddish spots on infected *Gomphrena* ◄ leaves.

265 Colletotrichum leaf spots are different sizes and cause distortion. ►

266 Tomato spotted wilt virus distorts new leaves, which also develop tan ◄ spots.

HEMEROCALLIS (DAYLILY)

267 *Armillaria* mycelia and rhizomorphs have formed on daylily rhizome.

268 Dead streaks from *Colletotrichum* form down the center of daylily leaf.

269 *Aureobasidium* causes reddish, diamond-shaped spots than can join to kill large portions of the leaf. ◄

270 *Colleocephalus* streak on daylily. ➤

HEUCHERA (CORAL BELLS)

271 Small brown spots are signs of anthracnose on *Heuchera*.

272 Pseudomonas leaf spot on *Heuchera* sometimes causes leaf distortion.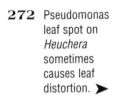

273 Foliar nematode on *Heuchera* appears as irregularly shaped dead spots with bright yellow borders.

HEUCHERA (CORAL BELLS)

274 Sunburn causes exposed leaves to bleach and ◄ die.

275 Large areas of a leaf can die from Botrytis leaf spot. ➤

276 Xanthomonas leaf spot on *Heuchera* causes small angular or circular spots with yellow ◄ borders.

HIBISCUS

277 *Choanephora* covers senescent hibiscus flowers with mycelia and spores.

278 Hibiscus rust forms on leaf undersides. Pustules can be seen on the upper surface as well.

279 Herbicides like 2,4-D cause leaves to turn yellow and become distorted.

280 Hibiscus leaves can sunburn when moved from shade to sun.

281 Molybdenum deficiency can distort hibiscus flowers. ◄

282 Molybdenum deficiency induces distortion, cupping, and stunting in hibiscus leaves. ◄

283 Leaves can turn yellow and drop from Xanthomonas leaf spot. ►

HIBISCUS

284 Xanthomonas leaf spot of hibiscus usually forms angular spots with bright yellow halos.

285 *Pseudomonas cichorii* spots on hibiscus are tan with purplish margins.

286 White flecks are the cast skins of
◄ aphids.

HOSTA

287 Grasshopper feeding creates large, irregularly shaped holes usually starting at leaf tips or
◄ edges.

288 Slugs feed all over leaves, especially ◀ between veins.

289 Botrytis leaf spot can create small brown spots anywhere on the leaf. ➤

290 Taxus weevils feed on leaf edges and leave semi-circular ◀ holes in the leaves.

291 Anthracnose produces large, irregularly shaped tan spots with brown borders that sometimes lose their centers.

292 Anthracnose can give leaves a tattered appearance.

293 Leaf margins that are brown or yellow are one sign of hosta water deficiency.

294 *Cercospora* produces brown spots of all shapes and sizes on hydrangea.

295 Mycosphaerella spots are roughly circular and surrounded by a purple or red ◄ border.

296 Wilt and plant collapse are signs of hydrangea root rot. ➤

HYDRANGEA

297 Sooty mold forms on leaves of plants infested with certain insects and mites.

298 Phyllosticta leaf spots are relatively large and surrounded with a black border as well as a reddish margin.

299 Botrytis blight on hydrangea makes flowers turn brown ◀ and collapse.

300 *Corynespora cassiicola* causes irregularly shaped reddish spots on hydrangea. ➤

HYDRANGEA

301 *Pseudomonas cichorii* spots on oakleaf hydrangea are vein-delimited and cause leaves to pucker.

HYPERICUM

302 Rust on *Hypericum* is caused by *Uromyces triquetus.* ◄

303 Pseudomonas leaf spot on *Hypericum* typically appears as angular brown areas. ►

IMPATIENS

304 Powdery mildew shows up as white spots on impatiens.

305 Cercospora leaf spot on impatiens forms in concentric rings and can coalesce.

306 Impatiens necrotic spot virus causes black ringspots with some tan areas on common impatiens.

IMPATIENS

307 Impatiens necrotic spot virus creates color breaks, leaf puckering, and distortion on New Guinea ◄ impatiens.

308 Tiny worms sometimes attack the impatiens shoot tips. ➤

309 *Pseudomonas syringae* often produces small, tan, circular areas with purplish margins.

310 Bleached white round spots on flowers may form when plants are watered during the hottest part of the day.

IMPATIENS

311 Alternaria leaf spots of common impatiens are small, roughly circular tan spots with a purplish margin.

312 Spider mites can create severe defoliation if not controlled when first seen.

313 New Guinea impatiens exposed to direct sunlight can develop sunburn. ◀

314 The yellowish-green strands and flowers of dodder (a parasitic plant) attack many garden flowers. ▶

Diseases of Annuals and Perennials   135

315 *Botrytis cinerea* infects leaves, covering them with fungal spores.

316 Botrytis blight can cause discrete leaf spots.

317 Botrytis blight on New Guinea impatiens flowers produces brown clear spots on ◄ petals.

318 Rhizoctonia crown rot collapse of common impatiens. ➤

IMPATIENS

319 Close-up of Rhizoctonia crown rot shows sunken stem lesions. ◀

320 Thrips feeding damage on common impatiens appears like scratch marks. ▶

IRIS

321 Iris rhizome with *Botrytis* spores.

322 Rust pustules on iris are brown rectangular patches.

323 An insect borer has caused leaf death.

324 Didymellina spot appears as elliptical brown areas ◄ with a greasy margin.

325 *Mycosphaerella* causes "fire" on bulbous iris. ➤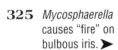

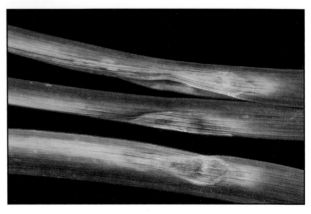

IRIS

326 Mycosphaerella leaf spot can appear as small elliptical reddish spots.

LIATRIS

327 When the apical bud is killed, the lateral buds form a witches'-broom.

328 Dieback on this *Liatris* was caused by a soil fungus.

LIATRIS

329 *Rhizoctonia* produces varying degrees of stunting due to root and bulb rot.

330 *Cercospora* causes spots that can combine across large areas ◄ of the leaves.

LILIUM (LILY)

331 Major portions of flowers can turn brown and collapse from *Sclerotinia.* ►

LILIUM (LILY)

332 *Sclerotinia* can cause small dark spots on unopened buds. ◄

333 "Fire" on lily is caused by *Botrytis elliptica*. ➤

334 An unidentified virus produces severe mosaic and distortion of lily buds and leaves. ◄

LIMONIUM (STATICE)

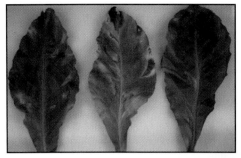

335 Foliar nematode infestation on statice appears as irregularly shaped areas sometimes with ◄ yellow borders.

336 *Botrytis* and several other fungi can blight flowers. ➤

337 Anthracnose may kill small ◄ statice in flats.

338 Close-up of Botrytis blight on statice flowers. ➤

LUPINUS (LUPINE)

339 *Colletotrichum* (anthracnose) can cause buds and shoot tips to die on infected lupine.

340 Close-up of brown to black stem lesion at soil line produced by anthracnose.

341 Powdery mildew on lower leaves of lupine. ➤

LYCHNIS

342 Septoria leaf spot on *Lychnis* can make leaves turn completely brown. ◄

343 Phyllosticta leaf spot on lower leaves of *Lychnis*. ➤

LYTHRUM

344 *Coniothyrium* creates cankers at bases of main stems and results in wilt and death. ◄

LYTHRUM

345 Coniothyrium canker in mid-portion of *Lythrum* ◄ stem.

346 *Rhizoctonia* can produce dead spots in pots of *Lythrum.* ➤

MATTHIOLA (STOCK)

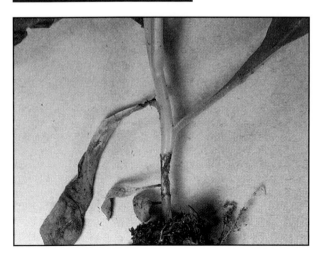

347 Basal stem canker on stock results from ◄ *Xanthomonas.*

Matthiola (Stock)

349 Close-up of Alternaria leaf spot showing black center from sporulation of the pathogen.

348 Alternaria leaf spot causes tan spots with black centers and yellow margins.

Monarda (Beebalm or Bergamot)

350 *Cercospora* creates small dryish tan spots on *Monarda.*

351 Powdery mildew can cause lower leaves to drop.

352 Large brown spots with a yellow border are signs of tomato spotted wilt virus infection. ◄

353 *Botrytis* infections on flowers are usually brown and show loss of tissue turgor. ➤

354 Pseudomonas leaf spot on *Nicotiana* appears along the veins. ◄

PACHYSANDRA

355 Herbicides can create stunted, distorted leaves.

357 *Rhizoctonia* infections in *Pachysandra* flats can produce large brown leaf spots as well as stem rot.

356 Cold winters can result in scorch on unprotected *Pachysandra*.

358 Volutella spot appears as circular brown areas that may turn the rest of the leaf yellow. ➤

PACHYSANDRA

359 Sunburn on exposed *Pachysandra* leaves.

PAEONIA (PEONY)

360 *Botrytis* can kill peony shoots.

361 Botrytis leaf spots on peony are irregularly shaped brown areas.

PELARGONIUM (GERANIUM)

362 Foliar nematodes form brown to black areas in geranium leaves. ◄

363 Downy mildew produces angular patches of yellow growth on leaf upper side. ►

364 This same mildew causes white growth patches on leaf underside. ◄

365 Bacterial leaf spot produced by *Xanthomonas campestris* pv. *pelargonii* is typified by small brown to purplish spots. ►

PELARGONIUM (GERANIUM)

366 Bacterial leaf spot caused by *Xanthomonas campestris* pv. *pelargonii* on *Geranium* spp. appears as small brown spots sometimes surrounded by ◄ yellow areas.

367 Bacterial leaf spot of florist geranium is typified by small brown spots surrounded by large yellow areas. ►

368 Pseudomonas leaf spot on geraniums creates brown spots that can ◄ grow together.

PELARGONIUM (GERANIUM)

369 Cucumber mosaic virus causes slight mosaic and distortion. ◄

370 *Alternaria* produces small, tan, roughly circular leaf spots on *Pelargonium graveolens*. ►

371 Geranium rust appears as small brown pustules mainly on leaf undersides. ◄

Pelargonium (Geranium)

372 *Pythium* causes black leg on geranium and appears as black watery areas on stems near the ◄ soil surface.

373 Oedema usually appears on geraniums when more water is available than is actually needed, especially when the weather is cool. ➤

374 *Corynebacterium fascians* causes distorted witches'-broom-like growth on geranium cuttings.

PELARGONIUM (GERANIUM)

375 Leaves can become twisted and distorted from some growth regulators.

◄

376 Botrytis blight can cause an entire flower head to become brown and matted with fungal mycelia and spores. ►

377 *Botrytis* can infect stems and petioles making them turn brown to black and sunken.
◄

378 Close-up of Botrytis blight on geranium flowers.

379 Sclerotinia blight occurs in centers of petunia and is typically accompanied by the pathogen's white mycelia and the black sclerotia.

380 Boron deficiency results in stunting and yellowing of the newest growth.

381 Phyllosticta leaf spot on petunia appears as sunken tan areas on lower leaves.

382 Impatiens necrotic spot virus infections can cause black ringspots.

PETUNIA

383 Botrytis blight on petunia flowers starts as white spots. ◄

PHLOX

384 Severe powdery mildew infection on phlox lower leaves. ◄

385 *Septoria* causes round spots of all sizes with a purple margin. ➤

PHLOX

386 Cercospora leaf spot starts as small brown areas with yellow margins. ◄

387 Foliar nematode infestation of plants on the right compared to healthy plants (left). ►

388 *Xanthomonas* infects phlox margins and invades tissues between leaf veins. ◄

PHLOX

389 Stem nematodes cause leaves to turn brown and die. ◄

390 *Sclerotium rolfsii* forms small, brown, mustard seed-sized sclerotia at stem bases. ➤

391 Fusarium cankers at phlox stem bases cause plants to wilt and die. ◄

PLATYCODON (BALLOON FLOWER)

392 Alternaria leaf spot on *Platycodon* appears first as small brown areas with yellow margins between leaf veins. ◀

393 Sclerotinia blight causes *Platycodon* to wilt and die due to stem cankers. ➤

POINSETTIA

394 Poinsettia bracts and leaves can develop large brown spots due to *Corynespora cassiicola.* ➤

395 Poinsettia bracts develop small yellowish spots when infected with *Xanthomonas campestris* pv. *poinsettiicola.*

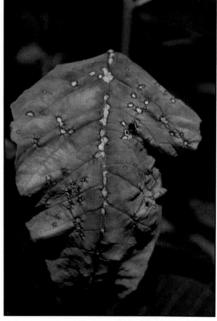

396 *Xanthomonas* causes small reddish angular spots surrounded by a yellow margin on poinsettia leaves.

397 Raised brown spots with yellow margins appear on leaves of poinsettia infected with scab.

POINSETTIA

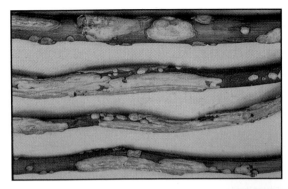

398 Scabby dry areas appear on stems and petioles of poinsettias with scab disease. ◄

399 *Botrytis* infections can start on leaf tips and cause watery areas that expand along the leaf veins. ➤

400 Powdery mildew forms on ◄ bracts and leaves.

POINSETTIA

401 *Choanephora* infects poinsettia flowers and can ◄ invade the stem.

402 Alternaria leaf spot starts as tiny brown areas that can distort the leaf development.

403 Rhizopus blight of poinsettia can kill plants if stems are infected. ➤

404 Downy mildew on potentilla is easiest seen on leaf undersides.

405 The orange pustules on leaf undersides are key symptoms of rust.

RHODODENDRON

407 Yellow spots on uppersides of leaves are sometimes signs of a scale infestation on undersides.

406 Rhododendron scorch is produced by desiccation during winter.

408 *Xanthomonas* sp. causes angular, dark brown water-soaked spots on rhododendron.

RHODODENDRON

409 Botryosphaeria canker results in dieback on infected rhododendron branches. ◀

410 Taxus weevil damage gives rhododendron leaves a serrated edge. ➤

411 *Discula* sp. causes large frogeye spots on rhododendron. ◀

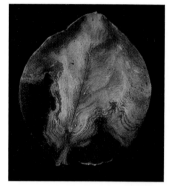

412 *Botrytis* can start at leaf
tips and slowly
encompass entire leaf.

413 Midge feeding
caused rhodo-
dendron tips to
become
distorted.

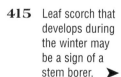

414 Root mealy-
bugs on a
rhododendron
◀ plant.

415 Leaf scorch that
develops during
the winter may
be a sign of a
stem borer. ➤

RHODODENDRON

416 Stem borer that was dissected from an ◄ infested twig.

417 *Mycosphaerella* creates brown, irregularly shaped leaf spots on rhododendron. ➤

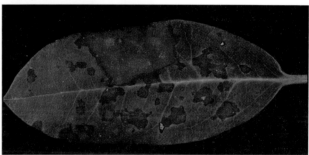

418 Upper portions of rhododendrons infected with *Phytophthora* are wilted.

RHODODENDRON

419 *Phytophthora* can cause brown streaks in the wood under the back of dead or dying stems.

420 Phyllosticta leaf spots are whitish on rhododendron.

421 Rhododendron rust is light tan or yellowish.

RHODODENDRON

422 Lacebugs cause leaves to have a speckled, grayish appearance.

ROSA (ROSE)

423 Downy mildew fungus can infect rose flowers and produce small red spots. ◄

424 Downy mildew fungus creates angular purplish spots on leaves. ►

Diseases of Annuals and Perennials 169

ROSA (ROSE)

425 Prunus necrotic ringspot virus causes light green mosaic in roses. ◄

426 Combined infections of prunus necrotic ringspot and apple mosaic viruses are common. ►

427 Rose mosaic virus makes bright yellow patterns. ◄

ROSA (ROSE)

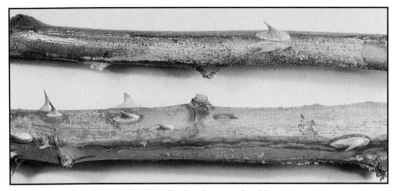

428 Coniothyrium canker usually develops on freshly cut rose stems.

429 Rose rust appears on leaf undersides in bright ◀ orange pustules.

430 Uppersides of rose leaves infected with rust have yellow spots. ➤

Rosa (rose)

431 Black spot on a Rugosa hybrid appears as tan spots with purplish borders. ◀

432 Black spot typically appears as black spots with diffuse margins surrounded by wide yellow halos. ▶

433 Spider mites injury on undersides of rose leaves. ◀

ROSA (ROSE)

435 Crown galls form at bases of roses infected with *Agrobacterium tumefaciens*.

434 Powdery mildew infections distort immature leaves.

436 Mycosphaerella leaf spot makes tan spots with discrete purple or red borders.

437 *Phomopsis* dieback starts at cut ends. The pathogen's light tan spores form in the canker area. ➤

RUDBECKIA (BLACK-EYED SUSAN)

438 Air pollution can cause upper leaf epidermis to turn white or tan.

439 *Pseudomonas cichorii* produces angular black spots on infected *Rudbeckia.*

RUDBECKIA (BLACK-EYED SUSAN)

440 Powdery mildew on *Rudbeckia*.

441 *Xanthomonas* sp. forms angular black spots on lower leaves.

RUDBECKIA (BLACK-EYED SUSAN)

442 Cercospora leaf spot on Goldstrum
◄ *Rudbeckia.*

443 *Botrytis* can infect entire crown
area of *Rudbeckia* when weather
is cool and wet.

SALVIA

444 Corynespora leaf spot on
Salvia makes leaves turn
◄ yellow and drop.

SALVIA

445 *Corynespora cassiicola* infects stems creating large black sunken lesions that kill small plants.

446 Alternaria leaf spot on Blue Victoria *Salvia* causes lower leaves to drop.

447 Typical *Botrytis* infection of *Salvia* flowers showing sporulation of the fungus. ➤

448 Marginal yellowing and browning are signs of desiccation.

449 Dark brown rust pustules on undersides of Blue Victoria *Salvia.*

450 Sweet potato whiteflies distort new leaves on Blue Victoria *Salvia.*

SALVIA

451 Macrophomina causes black cankers to
◄ form on red *Salvia*.

452 Powdery mildew spots on red *Salvia*.

STOKESIA

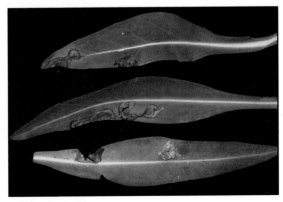

453 *Rhizoctonia solani* can cause discrete brown
spots on lower leaves of infected *Stokesia*.

STOKESIA

454 Alternaria leaf spots form as tiny brown or black areas on infected *Stokesia*.

455 Biden's mottle virus causes distortion and mottling.

456 Tomato spotted wilt virus creates distortion and blackening of new leaves.

STOKESIA

457 Colletotrichum leaf spots form primarily on leaf margins, and leaves frequently lose their centers.

SYRINGA (LILAC)

458 Leaf miner creates discolored tunnels in ◄ lilac leaves.

459 Older stages of lilac leaf miners roll the leaves over their bodies. ➤

Syringa (Lilac)

460 Close-up of powdery mildew on leaves.

Tagetes (Marigold)

461 Basal leaves of marigolds infected with *Fusarium* become matted and dead. ◀

462 Aster yellows disease causes infected marigolds to develop witches'-broom growth. ▶

TAGETES (MARIGOLD)

463 Powdery mildew on older leaves of marigold. ➤

464 Close-up of Botrytis blight on ◄ marigold flower.

465 Alternaria leaf spots are small, round and have a purple margin. ➤

466 Pythium root rot kills many marigolds in pots.

467 Thrips injury appears as small etched spots or flecks on marigold leaves.

TULIPA (TULIP)

468 *Botrytis* infects leaves during the cool, wet season.

469 *Botrytis* causes water-soaked flecks on infected tulip flowers.

VERONICA

470 Early signs of powdery mildew on *Veronica*.

Veronica

471 Advanced symptoms of powdery mildew on *Veronica*. ◄

472 Typical wounds caused by plant bug feeding on *Veronica*. ➤

Vinca (Periwinkle)

473 *Phyllosticta* causes vine death. ◄

VINCA (PERIWINKLE)

474 *Vinca* infested with cyclamen mites develop distorted, twisted, and stunted ◀ leaves.

VIOLA (PANSY)

475 Powdery mildew on pansy. ▶

476 Alternaria ◀ leaf spot on pansy.

VIOLA (PANSY)

477 Sclerotinia blight can produce collapse of plant.

478 Single spots from *Botrytis* can be circular and contain spores of the fungus.

479 Colletotrichum leaf spot on pansy appears as small circular spots with a purple border.

VIOLA (PANSY)

480 Black circular spots from Cercospora leaf spot sometimes join.

481 Cercospora leaf spot appears on lower leaves of pansy when moisture conditions are high.

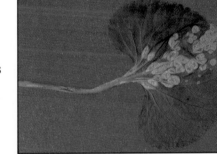

482 Rust on violet appears as corky yellowish spots. ➤

483 *Sphaceloma* causes scab on violet. ➤

VIOLA (PANSY)

484 Downy mildew sporulation can be seen as a grayish growth on the undersides of infected pansy leaves. ◄

ZINNIA

485 Xanthomonas leaf spot on zinnia is typified by angular brown spots that first form on lower leaves. ►

486 Powdery mildew disease on zinnia. ◄

ZINNIA

487 *Choanephora* can infect zinnia flowers causing them to turn brown and ◄ mushy.

488 *Alternaria zinniae* causes large reddish spots on zinnia. ➤

489 *Botrytis* infections on zinnia flowers give ◄ them a speckled look.

490 Aster yellows produces distorted, greenish zinnia flowers.

491 Close-up of *Cercospora* infections on a zinnia flower.

ZINNIA

492 Cercospora leaf spot on a young zinnia shows tan spots with purplish margins. ◄

493 Biden's mottle virus causes zinnia flowers and leaves to develop a mottle. ➤

494 Cold damage can make zinnia flowers dry out and die.

GLOSSARY

Adventitious bud — A bud that arises from a leaf, stem section between nodes, or roots rather than from the leaf axil.

Angular leaf spot — A spot with straight sides, bounded by leaf veins.

Anthracnose diseases — Certain diseases caused by *Colletotrichum, Discula,* and *Gloeosporium*; typified by leaf spots and blights.

Antibiotic — A complex chemical produced by a microbe that is toxic to another microbe.

Bacteria (sing. bacterium) — One-celled microorganisms that cannot be seen with the naked eye. Some bacteria cause plant diseases.

Bactericide — A pesticide that kills bacteria.

Biological control — Control of pests by using predators, parasites, and disease-producing organisms instead of chemical pesticides.

Blight — Extensive plant tissue death results when conditions are opportune for leaf spotting fungi and bacteria. Infection occurs throughout the plant and is not localized.

Canker — A dead area on a branch or stem that restricts development of the branch or plant beyond that point.

Chemical control — Control method based on using chemical pesticides or disinfectants to reduce or eliminate a pest population.

Chlorosis — The yellowing of a plant's normally green tissue.

Color breaking — Interruption in the normal color of petals or leaves with either darker or lighter areas.

Cortex — A cylinder of plant cells between the outer skin of the stem or root (epidermis) and the inner conductive (vascular) tissue (stele).

Crown gall — A common name for the plant disease caused by the bacterium *Agrobacterium tumefaciens*. The disease is typified by a gall at the plant's crown (base).

Cultural control — Methods that do not involve use of chemical or biological techniques to control pests. For example, sanitation is a cultural control.

Damping-off — Seedling collapse due to rot of seed before or after germination, generally due to fungal infection.

Desiccation — Drying of plant tissue due to lack of water.

Dieback — The progressive death of a shoot or branch tip caused by a root or stem disease, insect injury, nematode injury, or cold weather (to name a few).

Disease — Any disturbance of a plant that continuously interferes with its normal structure, function, or economic value. Diseases can be caused by fungi, bacteria, viruses, and other similar organisms.

Disinfectant — An agent that kills or inactivates pathogenic organisms within the plant or plant part.

Disinfestant — An agent that kills or inactivates organisms present on the surface of the plant or plant part, tools, potting bench, or other structure associated with the plant.

DNA — Deoxyribonucleic acid is the principal, complex organic acid of the chromosome, representing the cell's hereditary material.

Downy mildew — A common name for a group of foliar plant diseases characterized by the "downy" growth of the causal fungus on the lower leaf epidermis.

Epidemic — An outbreak of a disease in a high percentage of the plant population.

Epidermis — The outer skin of the leaf, stem, or root.

Eradicant — A pesticide that kills the pest after it appears.

Fumigation — The action of disinfesting the soil, structures, or growing surfaces, usually through use of a volatile gas or liquid pesticide.

Fungi (sing. fungus) — Organisms (often microscopic) lacking chlorophyll. Some fungi cause rots, mold, and plant diseases. Mushrooms are another type of fungus.

Fungicide — A pesticide that kills fungi.

Gall — A swollen area on leaves, stems, or roots that may be caused by insects, mites, fungi, bacteria, or nematodes.

Herbicide — A pesticide used to kill or inhibit plant growth.

Host — The living plant or animal that a pest depends upon for survival for at least part of its life cycle.

Immune — A state of not being affected by a disease or poison (total resistance).

Infection — The establishment of a pathogen or parasite within the host resulting in disease.

Infestation — Pests that are found in an area or location where they are not wanted.

Insecticide — A pesticide used to control insects.

Integrated control — The use of more than one approach or method of pest control; includes cultural, biological, and chemical practices.

Leaf abscission — Leaf drop.

Mites — A tiny arthropod closely related to ticks. Adults have eight jointed legs, two body regions, but no antennae (feelers) or wings.

MLO — Mycoplasma-like organism. MLOs are similar to a bacterium but lack a cell wall. They cause symptoms more like those seen in virus-infected plants.

Mosaic — A pattern of light and dark areas in a leaf that is frequently a sign of a virus infection.

Mottle — Irregular light and dark colored areas on plant parts that are frequently caused by a virus.

Mycelium — The vegetative part of a fungus analogous to the leaves, stems, and roots of a plant; the hypha or mass of hypha that make up the fungus body.

Necrosis — Death of plant cells; usually the affected tissue turns brown.

Nematode — A worm-like organism that may feed on or in plants and is generally microscopic. Nematodes have many common names, including roundworms, threadworms, and eelworms.

Oedema — Blisters on leaf undersides caused by physiological stress and excess water in the tissue.

Parasite — An organism that lives and feeds in or on another plant or animal (known as the host) and obtains all or part of its nutrients from the host.

Pasteurization — The destruction of selected pests in soil or other growing media by use of heat or chemicals.

Pathogen — Any disease-producing organism.

Pedicel — The slender stalk or stem of a single flower within a cluster of flowers.

Pest — A unwanted organism (insect, mite, nematode, bacterium, fungus, virus, weed, etc.) that attacks food, fiber, or ornamental crops desired by people.

Phloem — The part of a plant's conductive (vascular) system composed of those cells responsible for food transport within the plant.

pH — A measurement that expresses the acidity or alkalinity of a solution. A pH of 7, the value for pure distilled water, is regarded as neutral; a pH of 1 to 7 is in the acid range; and a pH of 7 to 14 is in the alkaline range.

Phytotoxic — Harmful (poisonous or injurious) to plant life.

Pith — The stem's central tissue.

Plasmid — A circular piece of DNA that is self-reproducing but not part of the hereditary makeup of the host or plant.

Powdery mildew — A common name for a group of diseases on foliage, characterized by a powdery white or gray growth of the causal fungus.

Protectant — A pesticide applied to protect plant surface before infection occurs by a pathogen.

Pustule — A blister-like fruiting structure of a rust fungus usually found on leaf undersides. When the structure ruptures, the colored reproductive spores are revealed.

Resistant plant — One that is not affected by a pest to the degree expected.

Ring spot — A spot formed by concentric rings of tissue that are colored differently than normal tissue; usually caused by a virus infection.

RNA — Ribonucleic acid, a component of the cell sap and nucleus of plants and animals; represents a small portion of the chromosomes of all cells.

Rust — A common name for a group of diseases caused by fungi; named for the colored spores formed in pustules on infected plant parts.

Sanitation — Any activity to eliminate or reduce the numbers of a pest present in a given area. Includes removal of diseased or infested plants or plant parts, weeds, and dead plant material.

Saprophyte — A organism that can live on dead or decaying organic matter.

Sclerotia (sing. sclerotium) — Resistant, seedlike structures formed by some fungi that overwinter or remain dormant until conditions are optimal for infection.

Senescence — Normal decline and death of plant tissues due to aging.

Solarization — A nonchemical means to kill pathogens, insects, and weeds in soil using a clear plastic tarp cover and the heat generated by sunlight over four to six weeks.

Spore — A fungus structure, analogous to the seed of a plant, that serves to reproduce and spread the fungus.

Sporulation — Spore production by a fungus.

Stele — A plant's conductive tissues composed of phloem and xylem cells in the root system.

Susceptible plant — Any plant capable of being injured or diseased by a pest; not resistant or immune.

Systemic — A pesticide that is taken up by one part of a plant and translocated to another part where it acts against a pest.

Tolerant — The state of being only slightly affected by a disease (partial resistance).

Toxicity — How poisonous a pesticide is to a organism; the ability of a pesticide to produce injury.

Vascular infection — Infection of the xylem or phloem of a plant by a fungus, bacterium, or other pathogen that can result in decreased water transport, wilting, and stunting.

Vascular tissue — A plant's conductive tissues composed of phloem and xylem cells.

Vector — A carrier of a disease-producing organism (pathogen); an insect or other animal that transmits a pathogen.

Virescent — Abnormal green color seen in flowers, resulting from disease such as aster yellows.

Virus — A submicroscopic pathogen (disease-producing organism) that needs living cells to grow and that can cause disease in plants. Viruses are too small to be seen with a normal microscope.

Water-soaking — Dark, greasy, or wet-appearing tissue surrounding leaf spots, caused by bacteria and sometimes fungi.

Witches'-broom — A disease symptom that appears as a cluster of small, weak shoots emerging from the same point on a stem or branch; caused by bacteria, fungi, MLOs, mites, and other pests.

Xylem — Nonliving cells in plants that conduct water upwards from the roots to the leaves.

REFERENCES[1]

Diseases of annuals and perennials

Coyier, Duane L., and Martha K. Roane. 1986. *Compendium of rhododendron and azalea diseases.* St. Paul, Minn.: American Phytopathological Society Press.

Daughtrey, M.L., and M. Semel. 1987. *Herbaceous perennials: disease and insect pests.* Cornell Cooperative Extension Information Bulletin 207.

Daughtrey, Margery, and A.R. Chase. 1992. *Ball field guide to diseases of greenhouse ornamentals.* Batavia, Ill.: Ball Publishing.

Ecke Jr., Paul, O.A. Matkin, and David E. Hartley. 1990. *The poinsettia manual.* 3rd ed. Encinitas, Calif.: Paul Ecke Poinsettias.

Forsberg, Junius L. 1975. *Diseases of ornamental plants.* Urbana, Ill.: University of Illinois Press.

Horst, R. Kenneth. 1990. *Westcott's plant disease handbook.* 5th ed. New York: Van Nostrand Reinhold Company.

Jones, Ronald K., and Robert C. Lambe. 1982. *Diseases of woody ornamental plants and their control in nurseries.* Department of Agricultural Communications, North Carolina State University.

Keim, Randolph, and Wesley A. Humphrey. 1987. *Diagnosing ornamental plant diseases.* An illustrated handbook. Univ. of Calif., Div. of Agric. & Nat. Res.

Pettingill, Amos. 1971. *The White-Flower Farm garden book.* New York: Alfred A. Knopf.

Powell, Charles C., and Richard K. Lindquist. 1992. *Ball pest & disease manual: disease, insect and mite control on flower and foliage crops.* Batavia, Ill.: Ball Publishing.

Sinclair, Wayne A., Howard H. Lyon, and Warren T. Johnson. 1987. *Diseases of trees and shrubs.* Ithaca, N.Y.: Cornell University Press.

Smith, Michael D. 1989. *The Ortho problem solver.* 3rd ed. San Ramon, Calif.: Ortho Information Services.

Strider, David L., ed. 1985. *Diseases of floral crops.* Vols. 1 and 2. Westport, Conn.: Praeger Scientific.

Insects and other pests of annuals and perennials

Carter, Cathy Cameron, K.F. Horn, and J.R. Baker. 1978. *Insect and related pests of flowers and foliage plants: some important, common, and potential pests in North Carolina.* The North Carolina Agricultural Extension Services. AG-136.

Daughtrey, M.L., and M. Semel. 1987. *Herbaceous perennials: disease and insect pests.* Cornell Cooperative Extension Information Bulletin 207.

---

[1] All publications eventually go out of print. When this occurs, a copy can usually be obtained through interlibrary loan by your local public library.

Johnson, Warren T., and Howard H. Lyon. 1991. *Insects that feed on trees and shrubs.* 2nd ed. Ithaca, N.Y.: Cornell University Press.

Klass, Carolyn, and Diane M. Karasevicz. 1993. *1993-1994 guide to pest management around the home.* Cornell Cooperative Extension Miscellaneous Bulletin 74.

Powell, Charles C., and Richard K. Lindquist. 1992. *Ball pest & disease manual: disease, insect and mite control on flower and foliage crops.* Batavia, Ill.: Ball Publishing.

Smith, Michael D. 1989. *The Ortho problem solver:* 3rd ed. San Ramon, Calif.: Ortho Information Services.

Chemical control for diseases of annuals and perennials

McCain, A.H., and A.O. Paulus. 1988. *Diseases of field-grown flowers.* 1988. University of California.

Perry, Leonard P., and Gary Dezeil. 1988. New England greenhouse pest control recommendations. New England Greenhouse Conference.

Smith, Michael D. 1989. *The Ortho problem solver.* 3rd ed. San Ramon, Calif.: Ortho Information Services.

Tayama, Harry K., ed. 1991. *Floriculture crops: Chemical use handbook: a guide for insecticide, miticide, fungicide, growth regulator, and herbicide application.* Ohio Florists' Association Bulletin No. 735.

Selecting and growing annuals and perennials

The Time-Life gardener's guide — perennials. 1988. Alexandria, Va.: Time-Life Books Inc.

Brickell, Christopher. 1989. *The gardener's encyclopedia of plants and flowers.* London: Darling Kindersley Ltd.

Broschat, Timothy K., and Alan W. Meerow. 1991. *Betrock's guide to Florida landscape plants.* Pembrook Pines, Fla.: Betrock Information Systems Inc.

Coleman, Helmer, M. Jane, and Karla S. Decker. 1991. *Pictorial guide to perennials.* Kalamazoo, Mich.: Merchants Publishing Co.

Gorkin, Nancy Kline, ed. 1988. Perennials: a nursery source manual. *Plants & Garden Brooklyn Botanic Garden Record* Vol. 44(4) Handbook #118.

Heriteau, Jacqueline. 1990. *The National Arboretum book of outstanding garden plants.* New York: Simon and Schuster.

Moggi, Guido, and Luciano Giugnolini. 1983. *Simon and Schuster's guide to garden flowers.* New York: Simon and Schuster.

Proctor, Rob. 1992. Annuals: a gardener's guide. *Plants and Gardens Brooklyn Botanical Garden Record* Vol. 48(4) Handbook #133.

Tayama, Harry, K., ed. 1989. Tips on growing potted perennials and biennials. The Ohio State University.

Woods, Christopher. 1991. Perennials: a gardener's guide. *Plants and Gardens Brooklyn Botanical Garden Record* Vol. 47(3) Handbook #128.

Wright, Michael. 1984. *The complete handbook of garden plants.* New York: Facts on File Publications.

INDEX

Acremonium wilt. Plant with:
Chrysanthemum 27, 92
Agrobacterium radiobacter 24, 84
Agrobacterium tumefaciens K84 23
annuals and perennials susceptible to
53
plant with: Rosa 173
Air pollution. See Sulphur dioxide
Algal leaf spot. Plant with: Camellia 82
Alternaria 21
damping-off. Plant with: Eustoma 22
leaf spot. Plants with: Bergenia 78;
Calendula 80; Chrysanthemum 90;
Coreopsis 98; Dahlia 101; Dianthus
106; Impatiens 135; Matthiola 146;
Pelargonium 152; Platycodon 159;
poinsettia 162; Salvia 177; Stokesia
180; Tagetes 183; Catharanthus 86,
87; Viola 187
plants with: Aquilegia 65; Zinnia 5
Alternaria zinniae 191
Anthracnose. Plants with: Aster 66;
Bergenia 78; Camellia 82;
Hemerocallis 3; Heuchera 123; Hosta
129; Limonium 142; Lupinus 143
Aphelenchoides. Plants with: Aquilegia,
Bergenia, Chrysanthemum, Heuchera
34
Aphids 31. Plant with: Hibiscus 127
Apple mosaic virus. Plant with: Rosa 170
Armillaria. Plant with: Hemerocallis 122
Ascochyta
leaf spot. Plants with: Aquilegia 63;
Clematis 97; Delphinium 104
petal blight. Plants with:
Chrysanthemum 95; Delphinium 104
wilt. Plant with: Clematis 97
Aster yellows disease 31
control strategies 44
plants with: Aster 67; Bellis 31;
Calendula 81; Callistephus,
Campanula, Catharanthus,
Chrysanthemum 31; Coreopsis 31,
99; Cosmos 99; Delphinium 31;
Gaillardia 31, 113; Petunia,
Rudbeckia, Salvia, Tagetes, Viola 31;
Tagetes 182; Zinnia 192
Aureobasidium. Plant with: Hemerocallis
122

Bacterial leaf spots and blights 4–7. See
also Xanthomonas campestris pv.
pelargonii
control strategies 43
plants with: Clivia 5; Begonia 77;
Delphinium 6
Biden's mottle virus. Plants with: Stokesia
180; Zinnia 193
Birds' nest fungi. See Saprophytes
Black spot. Plant with: Rosa 172
Boron deficiency. See Nutrient deficiencies
Botryosphaeria, canker. Plant with:
Rhododendron 25, 165

Botrytis 2, 4, 21
blight 2, 4. Plants with: Ageratum 57;
Begonia 76; Celosia 89;
Chrysanthemum 93; Crossandra 100;
Echinacea 109; Gerbera 116;
Hydrangea 131; Impatiens 136;
Limonium 142; Paeonia 2, 3;
Pelargonium 154; Petunia 156;
Tagetes 183
fungal cankers 25
leaf spot. Plants with: Fuchsia 112;
Heuchera 124; Hosta 128; Paeonia
149; Viola 188
plants with: Aquilegia 65; azalea 74;
Chrysanthemum 91; Cosmos 99;
Eustoma 110; Iris 137; Nicotiana
147; Paeonia 149; Pelargonium 154;
poinsettia 161; Rhododendron 166;
Rudbeckia 176; Salvia 177; Tulipa
185; Zinnia 3, 191
Botrytis cinerea 11. Plants with: Begonia
77; Impatiens 136

Calendula smut. Plant with: Calendula 80
Camellia variegation virus 82
Canker 15, 18. See also Botryosphaeria,
Coniothyrium, Erwinia amylovora,
Fusarium, Macrophomina,
Phytophthora, Sclerotinia,
Xanthomonas
control strategies 44
and dieback diseases 24–26
and fusarium. Plant with: Celosia 89
plant with: poinsettia 17
stem. Plants with: Aster 68;
Chrysanthemum 96; Forsythia 111
Cercospora
flower infection. Plant with: Zinnia 192
leaf spot. Plants with: Antirrhinum 62;
azalea 70; Coreopsis 98; Hydrangea
130; Impatiens 133; Liatris 140;
Monarda 146; Phlox 157; Rudbeckia
176; Viola 189; Zinnia 193
Cercosporella leaf spot. Plant with: Alcea
58
Chimera. See Natural sport
Choanephora. Plants with: Hibiscus 125;
poinsettia 162; Zinnia 191
Cold injury. Plants with: azalea 73; Zinnia
193
Colleocephalus. Plant with: Hemerocallis
122
Colletotrichum
leaf spot. Plants with: Antirrhinum 61;
Bergenia 4; Camellia 83; Gomphrena
121; Hemerocallis 122; Hosta 4;
Stokesia 181; Viola 188
petal blight. Plant with: Eustoma 110
plant with: Lupinus 143
Coniothyrium
canker, dieback 25. Plants with:
Lythrum 144, 145; Rosa 33, 171
leaf spot. Plant with: Yucca 3

Copper deficiency. See Nutrient deficiencies
Corynebacterium fascians. Plants with:
Chrysanthemum 90; Pelargonium
153
Corynespora cassiicola 4
annuals and perennials susceptible to
45
plants with: Hydrangea 131; poinsettia
159; Salvia 177
Corynespora leaf spot. Plant with: Salvia
176
Criconema 32
Crown gall 23–24
control strategies 44
plants with: Chrysanthemum 23; Dahlia
102; Rosa 24, 173
Crown rot. See Fusarium; Rhizoctonia;
Rot, root and crown
Cucumber mosaic virus 29, 31
annuals and perennials susceptible to
54
plants with: Aquilegia 31; Bougainvillea
80; Dahlia, Delphinium 31; Gladiolus
120; Pelargonium 152
Cyclamen mite. Plants with: Clematis 97;
Delphinium 104; Vinca 187

Damping-off 17, 19–23. See also
Alternaria, Pythium
control strategies 44
plant with: Antirrhinum 62
Desiccation 36 . Plants affected by:
Gardenia 115; Hosta 129;
Rhododendron 164; Salvia 178
Didymellina leaf spot 4. Plant with: Iris 4,
138
Dieback diseases 24–26
control strategies 44
Discula sp. Plant with: Rhododendron
165
Ditylenchus 32
Dodder. Plants with: Aster 65; Impatiens
135
Drought 24, 34
marginal burning 34, 35
plants with: Astilbe 69; toad lily 35
wilt 34

Edema. See Moisture, excessive
Erwinia spp.
amylovora , canker 25
annuals and perennials susceptible to
46–47
plant with: Chrysanthemum 91
Exobasidium fungus. Plant with: Camellia
83
Exobasidium vaccini. See Leaf gall fungus

Fertilizer deficiency. Plant with: Gloxinia
36
Foliar diseases 2–13; See also Nematodes
control strategies 43